Student Solutions Manual

to accompany

Reconceptualizing Mathematics for Elementary School Teachers

Second Edition

Judith Sowder
Larry Sowder
Susan Nickerson

W. H. Freeman and Company
A Macmillan Higher Education Company
New York

ISBN-13: 978-1-4641-0899-0
ISBN-10: 1-4641-0899-4

Printed in the United States of America

First printing

W.H. Freeman and Company
41 Madison Avenue
New York, NY 10010
Houndmills, Basingstoke
RG21 6XS England

www.whfreeman.com

Contents

Solutions to Supplementary Learning Exercises

Note to Students: The selected answers below give you some indication about whether you are on the right track. They are NOT intended to be the type of answers you would turn in to an instructor. **Your complete answers should contain all the work toward obtaining an answer**, as described in the Message to Students.

Examples of Complete Answers:

Learning Exercise 2a in Section 1.2

The quantities and their values are:

Regular price of CDs	$9.95
Regular price of tapes	$6.95
Discount this month	10%
Discounted price of CDs	90% of $9.95 = $8.96
Discounted price of tapes	90% of $6.95 = $6.26
New discount on 3 items	20%
Number of CDs bought	1
Number of tapes bought	1
Amount spent on CDs	$8.96
Amount spent on tapes	$6.26
Sales tax	6%
Amount spent	?

This drawing represents the problem.

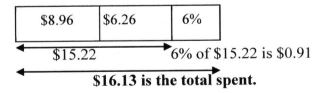

Learning Exercise 9 in Section 3.1

We need to find the weight of the sum of the medicine available from companies A, B, and C. A diagram will help. Here is one possible diagram.

```
                        1.3 mg
A      ————————————————————————————————

       (1.3 − 0.9) mg = 0.4 mg        0.9 mg
C      ——————————————|·····································

          0.4 mg        (0.5 × 1.3) mg = 0.65 mg
B      ——————————————·····························
```

Company A's medicine is represented by a line marked 1.3 mg. Company C's medicine is represented next because it is easier to find—it is 0.9 mg less than A's medicine. Thus, Company C has 0.4 mg of medicine. The difference between Company B's medicine and Company C's medicine is half of 1.3 mg, so it is 0.65 mg over and above Company C's medicine, which is 0.4 mg. Thus, Company B has 0.4 mg + 0.65 mg which is 1.05 mg. The total medicine furnished by the companies is (1.3 + 0.4 + 1.05) mg, or 2.75 mg.

Learning Exercise 2 in Section 5.4

$3 \times 10^4 \times 4 \times 10^6 = 12 \times 10^{10}$. This expression is not in scientific notation because 12 > 10. In scientific notation, this product would be expressed as 1.2×10^{11}.

Part I: Reasoning About Numbers and Quantities

Answers for Chapter 1: Reasoning About Quantities

Supplementary Learning Exercises for Section 1.2

1.   Finding a unit with which to measure "prettiness" and with which everyone agreed would be extremely difficult.

2. a. Quantities: Walt's height, Sheila's height, Tammi's height, and possibly the differences in pairs of the three people's heights, and even the difference in those differences. A drawing (for example, line segments, thin rectangles) should show the relative sizes and be labeled.

a.
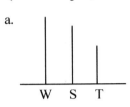

b.
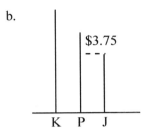

c.
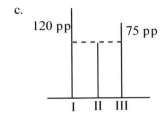

   b. Quantities: Pedro's salary, Jim's salary, Kay's salary, the difference in Pedro's and Jim's salary, etc. Your drawing should have labels and show the known $3.75 value for the difference in Pedro's and Jim's salaries.

   c. Quantities: The number of pages in each of the three books and the differences in page numbers (first and second, second and third, first and third). One could even consider the difference in the differences if that was appropriate for a particular question.

3.   Acting it out, a drawing showing different times, or a table for the runners' positions after various numbers of minutes can be helpful in seeing that Les catches up by 25 meters every minute, so she can catch up in 800 ÷ 25 = 32 minutes (if the race is not over). In 32 minutes, Les would have covered 32 × 250 = 8000 meters, so the race is not over. Les will take 10,000 ÷ 250 = 40 minutes to make the run. In 40 minutes, Pat will have covered 40 × 225 = 9000 meters, plus the 800 meters she was ahead at the start gives 9800 meters for Pat after 40 minutes. So Les is 10,000 − 9800 = 200 meters ahead when she wins.

4. a. Even an abstract drawing like the one to the right could be helpful. The entire drawing could be repeated until there are 24 B's.

   C C C C C C C C for
   B B B B B B

   b. Your drawing could focus on the 12 pictures, with $4.80 being associated with the whole group of 12, and then 35¢ associated with each picture.

5. Height, width, number of rooms, size of rooms, size of entrance hall, cost, …

6. Points scored, number of points more than (or less!) than the opponent, number of errors, number of rebounds, number of shots made, number of free-throws made, number of fouls,… Would your answers change if it had been the football team?

7. a. A drawing definitely helps most people. Here is one possibility.

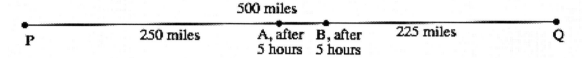

The trains have traveled $250 + 225 = 475$ of the 500 miles, so the locomotives are $500 - 475 = 25$ miles apart.

b. 260 miles. Train A has traveled 400 miles, and Train B has traveled 360 miles. So the trains have passed each other, as a drawing shows.

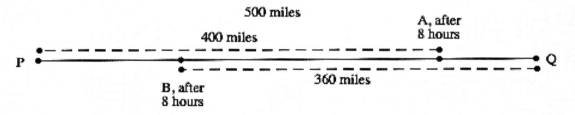

There are different ways of reasoning, but one way is to note that B is 140 miles from P, whereas A is 400 miles from P. Hence, the two locomotives are $400 - 140 = 260$ miles apart.

8. a. 490 miles. Make a drawing and keep in mind that Train D is heading *away from* P.

b. 430 miles.

c. Train C will eventually catch Train D. It is catching up by 10 miles every hour, so it will catch up in 50 hours.

Supplementary Learning Exercises for Section 1.3

1. Perhaps a little finger

2. A car might be 5 meters long, and perhaps also a kite tail and a python.

3. a. A meter is 100 centimeters.   b. A centimeter is one-hundredth of a meter.

   c. A decimeter is 10 centimeters.   d. A centimeter is one-tenth of a decimeter.

   e. A kilometer is 1000 meters.   f. A meter is one-thousandth of a kilometer.

4. The Latin word *mille* means 1000. In the metric system, the labels for units smaller than 1 are suggested by Latin words: 1 millimeter = one-thousand**th** meter

5. a. cm   b. mi   c. kg

6. 2 m (smallest)   510 cm = 5.1 m   300 dm = 30 m (largest)

7. a. 5   b. 0.5 (or just .5. So that the decimal point is not overlooked, SI recommends that a "0" be written to the left of the decimal point: 0.5.)
    c. 340   d. 0.034   e. 0.2 (or 0.200)   f. 200 000

Answers for Chapter 2: Numeration Systems

Supplementary Learning Exercises for Section 2.2

1.   a. 1000   b. 53   c. 538,649   d. 67   e. 6794   f. 8   g. 824,753,298

2.   a. 8,500,000 (metric: 8 500 000)

   b. 92,000,000,000 (metric: 92 000 000 000). In Britain, a billion is a million million rather than the thousand million in the U.S. A British billion would be a trillion in the U.S.

   c. $6,450,000,000,000 (metric: $6 450 000 000 000)

3.   a. "Six hundred and twenty-three thousandths"

   b. "Six hundred twenty-three thousandths"   (Notice the "and" in part (a).)

   c. "Twenty and eight hundredths"   d. "Twenty-eight hundredths"

4.   8.7645 tens, 87.645 ones, 876.45 tenths, 8764.5 hundredths, 87645 thousandths, 876450 ten-thousandths, and 0.87645 hundreds

5.   0.5 ones, 5 tenths, 50 hundredths

6.   a. four hundred seven hundred-thousandths
   b. fifty-one and 5 ten-thousandths
   c. four hundred and 5 hundredths (Notice the use of "and" here.)
   d. four hundred five thousandths (No "and"—remember to use "and" only for the decimal point in a decimal greater than 1, as in parts b and c.)

7.   a. 3.2   b. 7.53   c. 2000.017   d. 2.017

Supplementary Learning Exercises for Section 2.3

1.   1, 2, 3, 4, 5, 10, 11, 12, 13, 14, 15, 20, 21, 22, 23, 24, 25, 30, 31, 32, 33, 34, 35, 40, 41, 42, 43, 44, 45, 50 51, 52, 53, 54, 55, 100. A third digit is needed at 100_{six}, which is the square of six, or thirty-six.

2.   111, 1000, 1001, 1010, 1011, 1100, all in base two.

3.   The digit 3 does not appear in base three. Eighteen = 200_{three}.

4.   First, write the base-ten values of each place value in the unfamiliar system. Then calculate how many of each place value is needed, starting with the largest needed.

   a. 2101_{five} b. 424_{eight} c. 100010100_{two}

5. a. 774   b. 1882   c. 649

6. 354, with the small cube as the unit; 35.4, with the long as the unit; 3.54, with the flat as the unit; 0.354, with the large cube as the unit (any two)

7. 35.4_{seven}, with base-ten value $26\frac{4}{7}$.

8. Notice that each has four place values, so all four sizes of the usual kit of blocks will be needed for each.

    a. With the small cube as the unit, 4 large cubes and 2 small cubes

    b. With the large cube as the unit, 4 large cubes, 3 flats, 2 longs, and 5 small cubes

    c. With the long as the unit, 7 large cubes, 5 flats, 2 longs, and 3 small cubes

9. In binary (base two), 10 = two.

10. a. b^4
    b. $3b^3 + 2$ (or $3b^3 + 0b^2 + 0b + 2$)
    c. $b^5 + 2b^3 + 3b$ (or $1b^5 + 2b^3 + 3b$, or $1b^5 + 0b^4 + 2b^3 + 0b^2 + 3b + 0$)

11. a. 2030_b b. 41023_b c. 301026_b

Supplementary Learning Exercises for Section 2.4

1. If you work right-to-left and record them in the numerical work, the actions with the cut-outs support exactly the usual record-keeping with the "standard" algorithm.

    a. 413_{five}

    b. 1030_{four} (Notice that the final answer involves a larger piece than is in the cut-outs. With 3D materials, it would be a big $4 \times 4 \times 4$ cube; with the 2D, it could be a strip of four sixteens.)

    c. 361_{seven}

    d. 10101_{two} (Notice that the final answer involves larger pieces than are in the cut-outs. With the 2D, you need, for example, a square made up of two strips of two fours for the sixteen.)

2. If you trade as necessary to work right-to-left and record the trades in the numerical work, the actions with the cut-outs support exactly the usual record-keeping with the "standard" U.S. algorithm. Notice also that you can check these by adding the difference to the subtrahend (the number being subtracted).

      a. 213_{four} b. 224_{five} c. 246_{seven} d. 101_{two}

3. Base seven: small squares (1s); strip (long) of seven small squares (seven, 7^1); square (flat) of seven of those strips (forty-nine, 7^2)

4. Different answers are possible.

5. a.   four entries   b.   nine entries

+	0	1
0	0	1
1	1	10

+	0	1	2
0	0	1	2
1	1	2	10
2	2	10	11

       c. forty-nine (seven rows, with seven in each row)   d. one hundred   e. n^2

Answers for Chapter 3: Understanding Whole Number Operations

Supplementary Learning Exercises for Section 3.1

1. Each value (120 pages, 75 pages) refers to an additive comparison.

2. a. The sample drawing to the right shows how the additive comparisons can lead to either subtraction (23.4 – 0.064 = 23.336 s for the winner) or addition (23.4 + 0.92 = 24.32 s for third place), depending on how the quantities are related. Often the scales for the values are off because the values are unknown at the start.

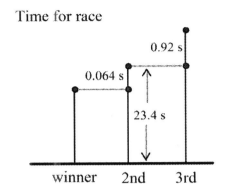

Time for race

b. The drawing suggests that if the difference is subtracted from the total, the result ($48.60) will equal twice the first trip's amount. So, the first trip's amount was $24.30, and the second $36.15.

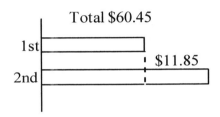

c. The possible drawing to the right shows that Will and Xavier have 49 cards, so Will has 49 – 23 = 26 cards.

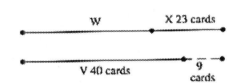

3. a. Compared: Nia and sister's collection vs. Dan and Ian's. Combined: Nia and her sister's collections; Dan and Ian's collections; the four collections

   b. Compared: Assets of Bank X vs. those of Bank Y; total of Banks X's and Y's assets vs. those of Bank Z. Combined: Assets of Bank X and Bank Y

4. There are many possibilities, so others may have different answers. Check that your values fit all parts of the stories.

5. Many possibilities. It is likely, however, that "how much more" or "how much less" appears in your story.

6. a. One possibility is to the right. It shows that Tia's current GPA is 2.7 and so Sue's current GPA is 2.7 + 0.3 = 3.0.

    b. Tia's current GPA and the 0.5 improvement are additively combined.

    c. Sue's and Tia's current GPAs are being additively compared.

Supplementary Learning Exercises for Section 3.2

1. You should have illustrations for the take-away, the comparison, and the missing-addend views. The illustrations should differ.

2. The differences is 83.5, the minuend is 165.8, and the subtrahend is 82.3.

3. a. Take-away   b. Missing-addend   c. Comparison

4. $y - z = x$

5. Here are examples, using marbles. There are many possibilities.
    a. Joe has 28 marbles, and Jean has 7. How many fewer marbles does Jean have than Joe? (Or, How many more marbles does Joe have than Jean?) Notice that two separate amounts are being compared in a how-much-more or how-much-less sense.

    b. Joe has 28 marbles. Only 7 of them are solid colors. How many of his marbles are not solid colors?

    c. Joe had 28 marbles but lost 7 of them in the park. How many marbles does Joe have now?

6. There are many possibilities. Here are examples.
    a. ...He had 36 feet of black irrigation pipe. How much less black pipe did he have? (The clearest earmark for a comparison subtraction situation occurs when two separate amounts are being compared in a how-much-more or how-much-less sense.)

    b. ...She needs 75 feet of pipe to finish the job. How many feet of pipe does she need?

    c. ...He used 29 feet for a flower bed. How many feet of pipe does he have now? (Clearest examples involve a taking-away, using, erasing, removing,... action.)

Supplementary Learning Exercises for Section 3.3

1. The child seems not to have any idea of what "+" really means.

2. These are sample directions for the arrows; different approaches could give different answers.

   a. Start at 189, go to 190, then to 192, then to 202, and finally to 232.

   b. Start at 936, go to 940, then to 1000, then to 1020, and finally to 1025.

   c. Start at 1000, go to 300, then to 270, then to 268.

   d. Start at 381, go back to 380, then to 320, and then to 319.

3. a. Probably $550 - 400$ b. $400 - 75$

4. Child 3 could start with 3 flats, 6 longs, and 4 small cubes. Starting on the right, trades of 1 long for 10 small cubes and then 1 flat for 10 longs would enable the take-aways. Child 6 would do similar trades, but in a different order.

5. No hints given! Be sure that your algorithm works for the second example.

Supplementary Learning Exercises for Section 3.4

1. To find the total when a number of same-sized quantities are combined (repeated-addition multiplication); to find the area of a rectangle or the number in an array (area or array multiplication); to find a fractional part of a quantity (part-of-an-amount multiplication); to find the number of possible occurrences of two or more events in succession (fundamental-counting principle).

2. With the repeated-addition view, 4×3 should be shown as 4 groups, each with 3 objects, and 3×4 as 3 groups, each with 4 objects. With the array/area view, 4×3 would have 4 rows with 3 in each row, and 3×4 would have 3 rows with 4 in each row. These would be the most common ways of showing 4×3 and 3×4.

3. Commutativity of multiplication ($4 \times 3 = 3 \times 4$)

4. Samples; be sure that your problem reflects the order of the factors (not in the written story as reported, but in the meaning for the operation).
   a. At $1.99 a bag, how much would Sean pay for 5 bags of chips?
   b. For a project, the teacher figures that she needs 12 pieces of construction paper for each of her 35 students. How many pieces of paper will she need?
   c. Dana bought 6 packages of dates, each of which weighed $\frac{3}{4}$ pound. How many pounds of dates did she buy?

   d. Eddie bought a six-pound bag of grapefruit and gave away $\frac{3}{4}$ of the bag to his relatives. How many pounds did he give away?

5. a. Your equation for associativity should follow the form $a \times (b \times c) = (a \times b) \times c$, or $(a \times b) \times c = a \times (b \times c)$. The two sides should calculate to give the same answer.
   b. Your equation for distributivity should be of one of the forms
   $a \times (b + c) = (a \times b) + (a \times c)$ OR $(a \times b) + (a \times c) = a \times (b + c)$ OR
   $(b + c) \times a = (b \times a) + (c \times a)$ OR $(b \times a) + (c \times a) = (b + c) \times a$. Again, the two sides should calculate to give the same answer.

6. a. Fundamental counting principle
   b. Repeated-addition multiplication
   c. Part-of-an-amount multiplication
   d. Array multiplication, or possibly repeated-addition multiplication

7. Part (a) should yield a tree diagram. Contrast your drawings for parts (b) and (c)—they should be different! Diagram for part (d) likely is an array.

8. a. $\frac{92}{113}$ is less than 1, so the product will be less than $\frac{578}{234}$.
   b. $\frac{578}{234}$ is greater than 1, so the product will be greater than $\frac{92}{113}$.

9. a. <u>multiplier</u> $\times$ <u>multiplicand</u> = <u>product</u>
   b. <u>factor</u> $\times$ <u>factor</u> = <u>product</u>

10. Any multiplication with a fraction less than 1.

11.
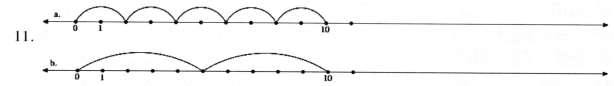

12. 4×3. Notice the order of the factors.

13. Apparently the student has not had experience with story problems involving the fundamental counting principle (or the student has forgotten it).

14. Only one was going to St. Ives! The others were going the opposite way. But other interpretations of the situation are possible, so if you use the riddle with children, you may hear other answers. See Wikipedia for the St. Ives riddle.

Supplementary Learning Exercises for Section 3.5

1. The repeated subtraction view entails removing or indicating groups of 2, giving three groups. The sharing equally view entails putting the 6 into 2 equal groups, giving 3 in each group.

2. a. and b. 1620 is the dividend, 36 the divisor, and 45 the quotient. Notice the different orders of the 1620 and the 36 in the two forms.

3. a. There will be 9 groups of 10, with 6 left over.

   b. There will be 9 in each of the 10 groups, with 6 left over.

4. a. No number makes sense. Removing 0 things from 6 things any number of times will never use up the 6.

   b. There are too many answers. Any number of removals of 0 things from 0 things makes sense.

   c. The related $n \times 0 = 6$ has no values for n that check.

   d. In the related $n \times 0 = 0$, *every* value for n checks.

5. Samples: check that yours do fit the indicated views.

   a. There are 150 children in Grades 5–6. How many teams of 5 children can be made for playground clean-up duty?

   b. There are 150 children in Grades 5–6. There are 5 buses available to take them on a field trip. How many children will be on each bus to keep the loads as equal as possible?

   c. The area of a rectangular banner that is 5 inches tall is 150 square inches. How long is the banner?

6. a. There are 23 fourteens in 322 (or 23 fourteens make 322).

   b. With 322, each of 14 equal shares will have 23.

   c. The solution to either $n \times 14 = 322$ or $14 \times n = 322$ is 23.

7. a. Sharing-equally division. $12 \times n = 168$ indicates that there are 12 groups of unknown size, totaling 168.

   b. Repeated-subtraction division. $n \times 12 = 168$ indicates that there is an unknown number of groups of size 12, totaling 168.

8. a. 500 seconds (from $93,000,000 \div 186,000$)

   b. $8\frac{1}{3}$ minutes (from $500 \div 60$)

   c. Both can be viewed as repeated-subtraction divisions.

9. There are different possibilities. Check to see that your finishes do fit the given ways of thinking about division.

10. a. The 15 meters is cut into 6 equal pieces.

    b. The 15 meters is cut into pieces each 2.5 meters long.

11. Do you have 3 equal pieces (for sharing-equally), or do you have pieces each 3 units long (for repeated-subtraction)?

12. It is praiseworthy that the child is trying to make sense of $0 \div 0$, but the child does not seem to be attending any meaning of division. You might say: "I see what you're thinking. But what does the division sign mean?"

13. You might ask the child how you can check a division by multiplying, say, $12 \div 6 = 2$ — probably easier to answer from the algorithm form $6\overline{)12}$. Then ask how the child could check $0 \div 0 = 100$.

14. If you are not sure, read Discussion 5.

15. 17, from $(12 \times 500) \div 350 = 17\ R\ 50$ (repeated-addition multiplication, repeated-subtraction division)

16. 19.2 miles (if you and your friend are along a straight line from you to the lightning). One way: You and your friend are 1.5 minutes apart. The sound will travel $768 \div 60 = 12.8$ miles in one minute. Then $1.5 \times 12.8 = 19.2$. Another way is to calculate how far each of you are from the lightning, and then subtract.

Supplementary Learning Exercises for Section 3.6

1. In the first division, the child apparently saw the $28 \div 4$ and passed over the tens place in the quotient. In the second division, the child calculated correctly but was mixing up the "R" format (R = 2 here) with the fraction form $\frac{remainder}{divisor}$.

2. It appears that the child is carrying over the algorithms from addition and subtraction in which the calculation is handled a place value at a time. So the child does 3×2 and 2×4, not realizing that the 3 also multiplies the 40 in the top factor and the 2(0) from the factor underneath also multiplies the 2 in the top factor.

3. a. 4×432 b. $520 \div 15$

4. Yes, the technique and its generalization are all right. The $560 \div 20$ might be seen as how many 2-tens are in 56-tens, but it may be easier to use what you may remember: $x \div y$ is equal to $\frac{x}{y}$ (this relationship is examined in Section 6.1). Then $(na) \div (nb) = \frac{na}{nb} = \frac{a}{b}$.

Answers for Chapter 4: Some Conventional Ways of Computing

Supplementary Learning Exercises for Section 4.1

1. Part (a) is not illustrated because the digit (0) for the flat place value is missing. The other parts are all right. Part (b) has the small cube as the unit. Part (c) has the long as the unit. Part (d) has the large cube as the unit. In part (e), the pieces could be base seven pieces.

2. Letting the small cube be the unit, you could show 1 flat, 3 longs, and 6 small cubes. Because it is to be take-away, you want to remove 8 longs and 9 small cubes. To support the usual algorithm, you would start with the small cubes. Because there are only 6 small cubes and you want to remove 9, you will have to trade 1 long for 10 small cubes (and this is exactly what you do symbolically), giving 1 flat, 2 longs, and 16 small cubes. Removing 9 small cubes leaves 7 small cubes. Dealing with the longs will involve another trade, getting 10 more longs for the flat (as in the algorithm). At this stage, there are 12 longs (and 7 small cubes), so you can remove 8 longs, leaving 4 longs and 7 small cubes, for 47.

3. (It is worthwhile to contrast how you worked Exercise 2.) You will start with two displays, one being 1 flat, 3 longs, 6 small cubes, and the other being 8 longs and 9 small cubes. To compare the small cubes first (to support the usual algorithm), you would again need to trade, as in exercise 2, and find that there are 7 more small cubes in the display for 136. Another trade of the flat for 10 longs allows you to compare the (now) 12 longs and 8 longs, showing that there are 4 more longs from the 136 display. Notice that the answer is perhaps less visible using the comparison method.

4. (It is worthwhile to contrast Exercises 2 and 3.) For the missing-addend view, you want to mimic 89 + ? = 136, so you will start with 8 longs and 9 small cubes. This start makes it difficult to lead to the usual algorithm, which shows that missing-addend is not a good starting point if you have in mind leading to, or supporting, the usual algorithm. Nonetheless, you can keep adding to the 8 longs and 9 small cubes, with trades necessary to get to the final 1 flat, 3 longs, 6 small cubes.

5. Using take-away and the long as the unit (why the long?), you would start with 5 flats, 2 longs, and 3 small cubes. Trades would enable you to remove 6 small cubes, 4 longs, and 3 flats, leaving 1 flat, 7 longs, and 7 small cubes, or 17.7.

6. Three groups, each with 2 longs and 4 small cubes, would allow you to focus on the 3 groups of 4 small cubes (the usual algorithm starts with the ones), trading 10 for a long and combining that with the 3 groups of 2 longs to get 7 longs and a product of 72. Notice that the 3 is not illustrated with 3 small cubes, but as the 3 groups.

7. You need only change the unit from the small cube to something that would allow hundredths. The flat would work for 3 × 0.24, and the large cube for 3 × 0.024. Otherwise, the display would be the same.

8.   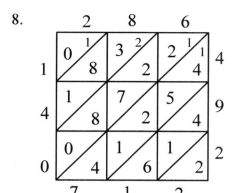   So, 286 × 492 = 140,712

9.   a. Think 64 = 2 × 32, so 64 × 95 = (2 × 32) × 95 = 2 × (32 × 95) = 2 × 3040 = 6080. (You would probably think, "Oh, just twice as much.")

      b. 9120   c. 12160   d. 95   e. Half as many, or 47.5   f. 190

10.   Each gives the same product, because the first factor has been halved and the second factor doubled, in effect multiplying by 1: $\frac{1}{2} \times 2 = 1$. Part (d) is easiest to do mentally.

Answers for Chapter 5: Using Numbers in Sensible Ways

Supplementary Learning Exercises for Section 5.1

1. a. 1567, seeing that 988 + 12 = 1000   b. 211, using 67 + 33, then 198 + (2 + 11)

    c. 635, left-to-right   d. 3111.6, just 29 – 18

    e. 3600   f. 1800, half as much as in part (e)

    g. 900, using (f), or every four 25s gives 100   h. 2700, 3 times as much as part (g)

    i. 1900 because nineteen 85s plus nineteen 15s is the same as nineteen (85 + 15)s— that is, the distributive property.

    j. 2300, because ninety-six 23s plus four 23s would be one hundred 23s.

2. a. 850   b. 1700   c. 32   d. 320   e. 8 × 32 = 256

    f. $13.20   g. $4.80   h. $7.50

Supplementary Learning Exercises for Section 5.2

1. a. 10.5% is about one-tenth, so about $15.

    b. About 5% of $50, or about $2.50

    c. 10% would be about $7.50, so about $\frac{3}{4}$ of $7.50. Something under $6.

    d. A 10% tip would be about $1.70, so 15% would be about $1.70 + 0.85, or about $2.50.

2. a. 35,000 (about 50 × 700)

    b. 33 × 40 is 66 × 20, or 1320. (You could add another 33, for 1353.)

    c. 4800 (about 8 × 600)

    d. $400 (about 2/3 of $600). If a closer estimate is needed, notice that $600 is about 15 more than $584.38, so since $\frac{2}{3}$ of 15 = 10, $390.

    e. A little less than $20 (about $\frac{1}{4}$ of $80)   f. $12 (one-third of $36)

Supplementary Learning Exercises for Section 5.3

1. About 400, using a class size of about 30

2. You can check how your benchmark worked by actually measuring.

3. 25 pounds or more, depending on the size of raccoons in your area

Supplementary Learning Exercises for Section 5.4

1.   The first factor is not ≥ 1 and < 10. 21.7×10^5 should be 2.17×10^6.

2.   a. 5,804,000   b. 8,000,000,000   c. 186,000   d. 3,900,000,000,000

      e. 0.0175   f. 0.00002703   g. 0.000000074   h. 0.0052368

3   a. 2.6829×10^{13} b. 8.236×10^3 c. 9.45×10^8

      d. $4.18923722 \times 10^{-1}$ e. 2.3×10^{-1} f. 4.85×10^{-4}

4.   1 thousand = 10^3; 1 million = 10^6; 1 billion (US) = 10^9; 1 trillion (US) = 10^{12}

5.   a. $26 \times 10^9 = 2.6 \times 10^{10}$ b. 5×10^5 c. $26 \times 10^{12} = 2.6 \times 10^{13}$

6.   a. 4.2×10^{10}
      b. 6.8×10^{13}
      c. $23 \times 10^{10} = 2.3 \times 10 \times 10^{10} = 2.3 \times 10^{11}$
      d. 1 million = 10^6, so $5 \times 4 \times 10^{3+6} = 20 \times 10^9 = 2 \times 10 \times 10^9 = 2 \times 10^{10}$

7.   If your display allows 8 digits, the largest number (without scientific notation) is 99,999,999.

8.   a. 10^8 = 100 million, so 5×10^{-8} = five hundred-million<u>ths</u>, or 0.000 000 05
      b. 10^4 = 10 thousand, so 166×10^{-4} = one hundred sixty-six ten-thousand<u>ths</u>, or 0.0166
      c. 10^6 = 1 million, so 19×10^{-6} = nineteen million<u>ths</u>, or 0.000 019

Answers for Chapter 6: Meanings for Fractions

Supplementary Learning Exercises for Section 6.1

1. Be sure to mention the unit (what = 1). $\frac{5}{9}$ means the unit is thought of as cut into 9 equal pieces, and you are considering 5 of those pieces.

2. Parts (a) and (c) have 6 equal pieces, so they are all right. Parts (b) and (d) do not have 6 equal pieces.

3. $\frac{17}{5}$. Your drawing should show 3 wholes, each cut into 5 equal pieces, giving 15 equal pieces in all. Those 15, plus 2 more fifths in another whole, give the 17 fifths.

4. Showing $\frac{19}{4}$ involves making copies of a whole, each cut into 4 equal pieces, and shading until 19 pieces are shaded. Each 4 pieces makes up a whole, so the shading, or the calculation 19 ÷ 4 = 4 R 3, tells that four wholes will be involved, with three pieces, or three-fourths, left over. $\frac{19}{4} = 4\frac{3}{4}$

5. a. $3\frac{1}{7}$ b. $12\frac{13}{15}$ c. $14\frac{1}{4}$ d. $19\frac{19}{35}$ e. $1\frac{1}{384}$ f. $3\frac{194}{219}$

6. $\frac{x}{y}$ can be interpreted as $x \div y$. So the sharing equally meaning would be that $\frac{x}{y}$ tells how much is in each if x's are shared equally among y's. The repeated subtraction would be that $\frac{x}{y}$ tells how many y's make, or are in, x.

7. Fifths and tenths are easy to show, using a point in the "center" to give the divisions. But how to cut the region into three equal parts is not at all clear.

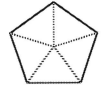 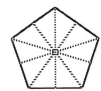

8. The unit for $\frac{1}{3}$ is the amount of money in the paycheck. The unit for $\frac{3}{4}$ is the amount spent on groceries. The unit for $\frac{1}{2}$ is the amount of space in the grocery cart. The unit for $\frac{4}{5}$ is the amount of money left from the paycheck (which is $\frac{2}{3}$ of the paycheck).

9. a. There is the same number of pieces (2) being considered in each, so the decision depends on which has larger pieces. Cutting a whole into 3 pieces gives larger pieces than does cutting it into 9 pieces, so $\frac{2}{3}$ is larger.

    b. Again, there is the same number of pieces (15) being considered in each, so the decision depends on which has larger pieces. A whole cut into 23 pieces will have smaller pieces than one cut into 22 pieces, so $\frac{15}{22}$ is larger.

c. The number of pieces in the whole is the same in each case (53 × 62), so the decision will depend on which fraction has more pieces. $2 \times 27 = 54$, but $3 \times 20 = 60$, so $\frac{3 \times 20}{53 \times 62}$ is larger.

d. $\frac{197}{192}$ is greater than 1, and $\frac{349}{360}$ is less than 1, so $\frac{197}{192}$ is larger.

10. a. The cake shown is 2 of 3 equal pieces, so cut it into 2 pieces (each will be $\frac{1}{3}$ of the cake). Tack on another such piece to get $\frac{3}{3}$ of the cake, the whole cake.

b. Make 2 and ¾ copies of the answer in (a).

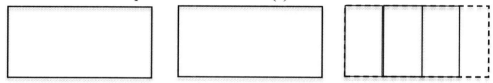

c. The whole cake would be worth $6.75, each third being worth $2.25.

11. One way: The distance between the given points is 1 unit. Cut that distance into 4 equal parts to get ¼ unit. For part (a), go to the left of the point for ¼ by one of those ¼ units. For (b), go to the right of the point for 1¼ three of the ¼ units. Finally for part (c) go another ¼ unit to the right of the point for 2.

12. $109.5 = 109\frac{1}{2} = 109\frac{4}{8}$, so the distance between the given points is 3/8 unit. Cut that distance into 3 equal parts to get 1/8 unit. Then for part (a) go up from the point for $109\frac{7}{8}$ by one of those 1/8 units, for part (b) go down four 1/8 units from the point for 109.5, and from that point, for part (c) go up two of the 1/8 units.

Supplementary Learning Exercises for Section 6.2

1. Equivalent fractions have the same value.

2.

a. 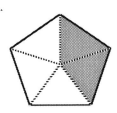   b.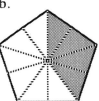

c. Show $\frac{6}{5}$, as in part (a), but using two pentagons. Then cut each fifth into two equal pieces, as in part (b).

3.   a. Samples (there are others): $\frac{5}{6}, \frac{10}{12}, \frac{15}{18}, \frac{20}{24}, \frac{25}{30}, \frac{30}{36}, \frac{35}{42}, \frac{40}{48}, \frac{45}{54}, \frac{50}{60}, \frac{500}{600}, \frac{5000}{6000}, \cdots$

b. The fractions are all equal to one another.

4.

a. 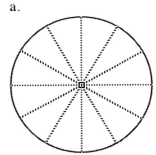   b.   Shade 9 pieces.

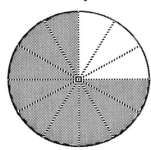

5.   a. $\frac{3}{4}$ b. $\frac{21}{32}$ c. $\frac{7}{40}$ d. $\frac{3}{4}$ e. $\frac{36}{125}$

f. $\frac{y}{z}$ g. $\frac{1}{150}$ h. $\frac{1}{ab^6}$ i. $\frac{5x}{6}$ j. $\frac{1}{x^4}$

6.   a.

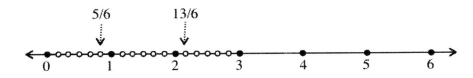

b.

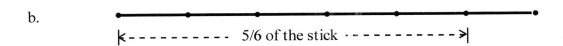

c. The units are different, with the unit on the number line the length between any two consecutive numbers, and the unit for the stick is the length of the whole stick.

7.   a. $\frac{163}{190}$ because it is larger than a half, and the other fraction is less than a half.

b. $\frac{29}{65}$ because $\frac{29}{65} = \frac{3 \times 29}{3 \times 65} = \frac{87}{195}$.

c. $\frac{44}{81}$ because $\frac{44}{81}$ is more than $\frac{3}{81} = \frac{1}{27}$ more than a half, but $\frac{37}{72}$ is only $\frac{1}{72}$ more than a half.

d. $\frac{83}{108}$, using a common denominator of 216.

8. $\frac{12}{15}$ of the marbles are black. In your diagram, to show 5 equal parts, put the marbles in clusters of 3 so that the black marbles are in 4 of the clusters.

9. Because $\frac{2}{3} = \frac{4}{6}$, the distance between the given points is 1/6 unit. One way: From the point for 2/3, go to the left four of the 1/6 units to get the point for $\frac{0}{6} = 0$. For part (b), go two of the 1/6 units to the right from the point for 5/6. For part (c), go two of the 1/6 units to the left of the point for 5/6.

10. Because $1\frac{1}{3} = 1\frac{4}{12}$, the distance between the given points of 5/12 unit. Cut that distance into five equal parts to get 1/12 unit. One way: Go to the left of the point for 11/12 by five of the 1/12 units to get to 6/12 (= ½). For part (b), go to the left of the point for 11/12 by two of the 1/12 units to get to the point for 9/12 (= ¾). For part (c), from the point for 9/12, go one 1/12 unit to the left to get to the point for 2/3 (= 8/12).

Supplementary Learning Exercises for Section 6.3

1. Each has a terminating decimal, because only factors of 2 and/or 5 appear in the denominators. The factor of 3 in the denominator of $\frac{699}{300}$ is not essential because $\frac{699}{300} = \frac{233}{100}$.

2. a. $\frac{3333}{10,000}$ b. $\frac{8374}{1000} = \frac{4187}{500}$ c. $\frac{7}{9}$ d. $\frac{63}{99} = \frac{7}{11}$

    e. $\frac{632}{999}$ f. $\frac{431.4}{99} = \frac{4314}{990} = \frac{2157}{495} = \frac{719}{165}$ (Your instructor might accept $\frac{4314}{990}$.)

    g. $\frac{154.73}{9} = \frac{15473}{900}$ h. $\frac{2299.2}{999} = \frac{22992}{9990} = \frac{11496}{4995} = \frac{3832}{1665}$ ($\frac{22992}{9990}$ may be enough.)

3. A rational number is any number whose decimal is either a terminating decimal or a repeating decimal. An irrational number is any number whose decimal neither terminates nor repeats. A real number is any number that can be represented by a decimal.

4. Whole number are rational numbers because they can be written as terminating decimals. For example, 173 = 173.0.

5. The decimal form of $\sqrt{5}$ goes on forever without any repeating block of digits.

6. π is indeed irrational. $\frac{22}{7}$ is only an approximate value for π.

7. Samples (There are actually infinitely many possibilities for each.)

    a. 0.985, 0.988   b. 1.0401, 1.0402,…   c. 0.49991, 0.49992,…

8. The equation is not true. 3.1416 cannot be *exactly* equal to π.

9. a. Half of an eighth, or half of 0.125. 0.0625, or 6.25%

     b. $3 \times 0.125 = 0.375$, or 37.5%

     c. Half of one-third, or half of $33\frac{1}{3}\%$. This is a slightly more difficult mental calculation, but half of 32 is 16, and then half of $\frac{4}{3}$ is $\frac{2}{3}$. So $\frac{1}{6} = 16\frac{2}{3}\%$. ($0.16\frac{2}{3}$, a mix of decimal and fraction within a single number, is usually avoided.)

     d. Half of $\frac{1}{6}$, or $8\frac{1}{3}\%$. Note the use of part (c).

     e. $\frac{11}{12}$ is $\frac{1}{12}$ less than 1, so $100 - 8\frac{1}{3} = 91\frac{2}{3}$ percent. Note the use of part (d).

Supplementary Learning Exercises for Section 6.4

1. Each addend is greater than $\frac{1}{2}$, so the sum should be more than $1\frac{1}{2}$. But the given answer is less than $1\frac{1}{2}$.

2. a. $\frac{1}{2}$ b. $\frac{1}{3}$ c. $\frac{3}{4}$ d. $\frac{2}{3}$ e. $\frac{2}{3}$

3. a. About $125, using $2\frac{1}{2} \times \$50$. The estimate is too small because each factor used is less than the corresponding factor in the original.

     b. About 18. The first two addends should give about 11, and the last two, about 7. The estimate is a little large because the 11 and 7 are a bit large.

     c. About 27, using $13\frac{1}{2} \div \frac{1}{2}$ and thinking, how many $\frac{1}{2}$ s are in $13\frac{1}{2}$? Because $\frac{8}{15}$ is larger than $\frac{1}{2}$ (and $13\frac{1}{2}$ is larger than $13\frac{7}{16}$), 27 is too many.

4. About $15

5. a. Larger: a numerator larger than 35 but close to 35. Smaller: a numerator smaller than 35 but close to 35.

     b. Larger: a numerator larger than 20 but close to 20. Smaller: a numerator smaller than 20 but close to 20.

     c. Larger: a numerator larger than 120 but close to 120. Smaller: a numerator smaller than 120 but close to 120.

6. The drawing was created to make the following percents and fractions; your estimates should be in the neighborhood of the following. (Note the use of Supplementary Learning Exercise 9 in Section 6.3.)

A and B each 25%, ¼;   C and D each $16\frac{2}{3}$%, $\frac{1}{6}$;   E and F each $8\frac{1}{3}$%, $\frac{1}{12}$. Your total for the percents should be about 100%, and for the fractions, about 1.

7. Somewhat more than 25%, by thinking of the 3.8 as a little less than 4.

8. 68% is about two-thirds, so cut the segment into two pieces, each one-third, and make the new segment two such pieces longer than the one given.

9. Candidate A received 59% of the vote, and B received 41%. Did you use a drawing, and the fact that the total vote would be 100%?

Answers for Chapter 7: Computing with Fractions

Supplementary Learning Exercises for Section 7.1

1. $88\frac{7}{8}$ inches. The 8 feet = 96 inches. Did you calculate $(96 - 4\frac{3}{4}) - 2\frac{3}{8}$? Or did you calculate $96 - (4\frac{3}{4} + 2\frac{3}{8})$? Do you see both ways of reasoning?

2. a. $17\frac{3}{4}$ miles, from $5\frac{1}{8} + 3\frac{3}{4} + 5\frac{1}{8} + 3\frac{3}{4}$

    b. $53\frac{1}{4}$ miles. Which did you do: Triple each length and then add, or triple the answer from part (a)? Your instructor may ask the name for the important property involved in assuring that the two methods will give equal answers:
    $3 \times (5\frac{1}{8} + 3\frac{3}{4} + 5\frac{1}{8} + 3\frac{3}{4}) = (3 \times 5\frac{1}{8}) + (3 \times 3\frac{3}{4}) + (3 \times 5\frac{1}{8}) + (3 \times 3\frac{3}{4})$.

3. $\frac{7}{12}$ gallon. Do you see that the results from the take-away actions could be determined by adding the amounts used, and then doing a take-away subtraction from the 2 gallons?
    $2 - \frac{1}{2} - \frac{2}{3} - \frac{1}{4} = 2 - (\frac{1}{2} + \frac{2}{3} + \frac{1}{4})$, or more generally, $a - b - c - d = a - (b + c + d)$

4. For part (a), make sure that the $\frac{3}{4}$ (quart of milk, say) is taken away from a $\frac{7}{8}$ (quart of milk), whereas for a clear-cut part (b), the two should be separate quantities. Part (c) should involve an addition situation that could be described by $\frac{3}{4} + n = \frac{7}{8}$.

5. a. $72\frac{1}{4}$ b. $128\frac{1}{8}$ c. $245\frac{2}{9}$ d. $999\frac{1}{18}$

6. a. $7\frac{5}{8} - 5\frac{7}{8} = 2 - \frac{2}{8} = 1 + \frac{8}{8} - \frac{2}{8} = 1\frac{6}{8}$ ($= 1\frac{3}{4}$ for us, and perhaps Killie)

    b. $10\frac{1}{3} - 6\frac{4}{9} = 10\frac{3}{9} - 6\frac{4}{9} = 4 - \frac{1}{9} = 3 + \frac{9}{9} - \frac{1}{9} = 3\frac{8}{9}$

    c. Probably not. because the subtraction does not require any renaming.

    d. By now you should appreciate that many times there may be more than the standard way of calculating. So it is to be hoped that you would not immediately rule out Killie's method.

7. $\frac{1}{3} + \frac{4}{9} + \frac{1}{6} = \frac{17}{18}$, so the rancher made an arithmetic mistake (or perhaps a horse had died). The classical solution is that the lawyer added his horse to the herd of 17, gave 6 to Curly, 8 to Buck, and 3 to Cindy Lou—and rode back to town on his horse!

8. a. $1\frac{23}{24}$ b. $2\frac{35}{48}$ c. $2\frac{25}{36}$

9. $5\frac{1}{4} - 2\frac{3}{4} = 2\frac{2}{4} = 2\frac{1}{2} = \frac{5}{2}$ so we can find ½ unit pieces by cutting the distance between the given points into five equal pieces, and ¼ unit pieces by cutting one of the ½ unit pieces into two equal pieces. For part (a), $6\frac{1}{2} - 5\frac{1}{4} = 1\frac{1}{4}$, so one way is to go the right from the point for 5¼ by two of the ½-unit spaces and one of the ¼ unit spaces (or by five of the ¼-unit

spaces). Part (b) is most easily accomplished by going to the left from the 2¾ point by the distance (2½) between the given points. For part (c), go to the left of the point for 5 ¼ by two of the ½ unit distances.

10. Using associativity and commutativity of addition allows us to rethink of the given problem in the following way: $(\frac{40}{99} + \frac{59}{99}) + (\frac{7}{3} + \frac{2}{3}) + \frac{5}{39}$, which in turn gives $\frac{99}{99} + \frac{9}{3} + \frac{5}{39} = 4\frac{5}{39}$.

Supplementary Learning Exercises for Section 7.2

1. Both $\frac{c}{d}$ and $\frac{e}{f}$ refer to the same unit.

2. a.

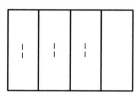

$\frac{3}{4}$ of the region

b.

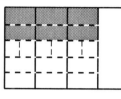

$\frac{2}{5}$ of the $\frac{3}{4}$

c.

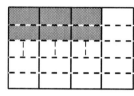

$\frac{2}{5}$ of $\frac{3}{4} = \frac{6}{20}$ of the region

b. There are two rows of 3 darkened regions (2×3), and the whole unit has been cut into 5 rows with 4 equal pieces in each row, or 5×4 pieces of the same size.

3. a. Ronnie paid $\frac{2}{9}$ (from the $\frac{4}{9} + (\frac{3}{4} \times \frac{4}{9}) = \frac{7}{9}$ paid by Ally and Bella). So each ninth of the cost is $400, and the whole cost would be $3600.

    b. Ally paid $1600, and Bella paid $1200.

4. a. $\frac{7}{24}$ b. $1\frac{1}{2}$ c. $\frac{1}{9}$ d. 4

    e. In multiplication, the numerator in the product and the denominator in the product are the result of multiplying the individual numerators and denominators. "Canceling" just recognizes that common factors in the numerators and denominators can be eliminated before multiplying the fractions rather than waiting until the product of the numerators and the product of the denominators are calculated.

5. $200. The first two payments took care of $\frac{2}{5} + (\frac{3}{4} \times \frac{2}{5}) = \frac{7}{10}$ of the loan. So the $60 represented $\frac{3}{10}$ of the loan, and each tenth of the loan was $20.

6. a. $100\% - 30\% = 70\%$ b. $100\% - x\%$

    c. About $60. One way is to calculate the discount first (25% of $79.95) and then subtract the discount from the regular price. Another way is to calculate 75% of $79.95.

7. There are many possibilities. Be sure, for example, that the 5, the 2/3, and the 3/4 apply to the amounts represented by the *second* factors, and that the unit for each of the second factors is clear (e.g., pounds or cups of sugar).

8. a.   The technique does not recognize that the first factor affects *all* of the second factor.

    b. $15\frac{5}{4} = 16\frac{1}{4}$

The most common algorithm taught is to change each of the factors to a fraction and multiply. (This algorithm could also be used on part (b), $5 \times \frac{13}{4} = \frac{65}{4} = 16\frac{1}{4}$, but is not necessary.)

    c. $\dfrac{10}{3} \times \dfrac{37}{5} = \dfrac{370}{15} = \ldots = 24\dfrac{2}{3}$ d. $\dfrac{127}{9} \times \dfrac{103}{8} = \dfrac{13081}{72} = \ldots = 181\dfrac{49}{72}$

9. a. One possible drawing:

    b. The 60 new ones must be the additional 20% over last year's, so 1% would be 3 employees, and 100% would be 300 employees last year.

    c. 300 + 60 = 360 employees this year

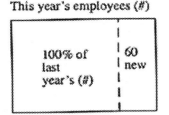

This year's employees (#)

10. a.   20%, the number in the whole student body; ¼, the number of voters; 35%, the number of voters

    b.   One possible drawing is to the right.

    c.   Is the drawing accurate enough to be trusted? We should check: A and B together have 60% of the vote, leaving 40% of the vote for C, so C does win.

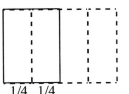

Number in student body

    d.   One way: C's 800 votes is 40% of the voters, so 1% of the voters would be $800 \div 40 = 20$ voters, and so 100% of the voters would be 2000 voters. But these 2000 voters are only 80% of the student body. 1% of the student body would be $2000 \div 80 = 25$ students, and so 100% of the student body would be 2500 students.

Supplementary Learning Exercises for Section 7.3

1. a. There are 6 one-fourths in $1\frac{1}{2}$.

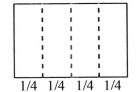

1/4 1/4 1/4 1/4   1/4 1/4

    b. Pieces i and ii give one $\frac{2}{3}$, and

pieces iii and iv give a second $\frac{2}{3}$.
With some added marks, piece v is $\frac{1}{4}$
of another $\frac{2}{3}$. Notice the care
necessary in interpreting piece v:
What part of another $\frac{2}{3}$ is piece v?

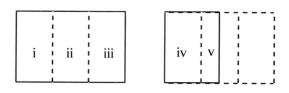

d. How many $1\frac{1}{2}$ s are in (or make)
1? There is not a whole $1\frac{1}{2}$ in 1;
only a part of a $1\frac{1}{2}$ will make 1. To
get equal parts in the $1\frac{1}{2}$, put in the
other half mark.

What part
of the gray

makes this?

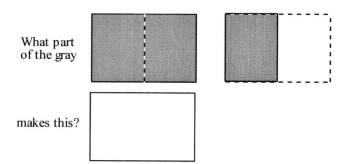

3. Showing the equal shares may involve
    putting in other cutting marks. For
    example, for part (a), each share is
    $\frac{1}{2} + \frac{1}{4} = \frac{3}{4}$.

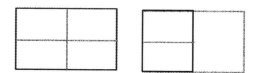

4. a and b. Each fraction refers to the same unit.

    c. The $\frac{c}{d}$ and the $\frac{e}{f}$ refer to the same unit, but the unit for the $\frac{a}{b}$ is the ($\frac{c}{d}$ of the unit).

    d. For the repeated-subtraction view, the first two fractions refer to the same unit, but the $\frac{e}{f}$
    tells how many of the $\frac{c}{d}$ s of the unit there are in the $\frac{a}{b}$ of the unit.

    (For the rarer—in U.S. curricula, for fraction divisors—sharing-equally view, the $\frac{a}{b}$ and the
    $\frac{e}{f}$ refer to the same unit, but the $\frac{c}{d}$ is some other unit.)

5. Notice how important knowing the units for the different $\frac{1}{3}$ s is.

    a. $6\frac{5}{12}$ cups   b. $4\frac{1}{2}$ cups

6. a. $\frac{4}{3}$ times   b. $2 \times \frac{4}{3}$ times (or $2\frac{2}{3}$ times)   c. $3 \times \frac{4}{3}$ times (or 4 times)

    d. $4\frac{1}{2} \times \frac{4}{3}$ times (or 6 times)   e. $\frac{7}{8} \times \frac{4}{3}$ times (or $1\frac{1}{6}$ times)

7.   What multiplication will "undo" a multiplication by $\frac{3}{4}$? Multiply by $\frac{4}{3}$, so set the machine at $133\frac{1}{3}$%, if possible, or if the machine allows only whole numbers, use 133% for a close approximation.

8.   Even without knowing the personality of the child, you can be sure that the child is not applying a meaning for division. You might ask what $6 \div 2 = 3$ tells you to see whether the child does have a "how many 2s are in 6" understanding. If so, ask what $\frac{3}{4} \div \frac{1}{4} = 3$ tells you.

9.   a. You can tell how many c's are in $a + b$ by seeing how many c's are in a, how many c's are in b, and adding those results. More algebraically,

$$(a+b) \div c = (a+b) \cdot \frac{1}{c} = (a \cdot \frac{1}{c}) + (b \cdot \frac{1}{c}) = (a \div c) + (b \div c).$$

b. It is not at all clear that you can tell how many $(b + c)$'s are in a, by seeing how many $(b + c)$'s are in a and then in b, and adding those results. If you avoid zeroes, you will have a "counterexample." For instance, $24 \div (2+3) \neq (24 \div 2) + (24 \div 3)$.

Answers for Chapter 8: Multiplicative Comparisons and Multiplicative Reasoning

Supplementary Learning Exercises for Sections 8.1 and 8.2

1. a. 5:3 or 5 to 3   b. $\frac{5}{12}$ c. $\frac{5}{3}$ or $1\frac{2}{3}$ d. $\frac{3}{5}$

   e. 8   f. 2   g. 8:4 or 8 to 4

2.

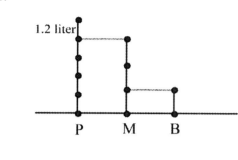

then   1.2 liter

(5x1.2 liters, or 6 liters)

(1.6 liters)

P   M   B

(4x1.2 liters, or 4.8 liters

So, the total consumption was 6 + 4.8 + 1.6 = 12.4 liters. How did your quantitative analysis and drawing help you?

3.

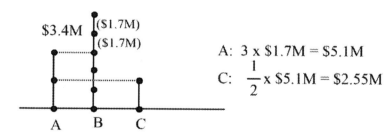

A: 3 x \$1.7M = \$5.1M

C: $\frac{1}{2}$ x \$5.1M = \$2.55M

What other quantities do you now know values for?

4. a.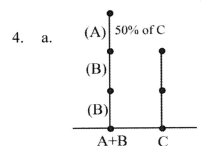

Total of the 5 pieces = 60 games
So each piece = 12 games.
A won 12 games, B 24, and C 24.

b.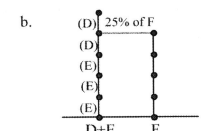

9 equal pieces -- 72 games
1 piece -- 8 games
D won 16 games, E 24,
and F 32.

5. One possible diagram:

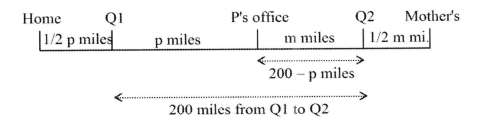

(Algebra could be used to solve this problem, but it is not difficult to find numbers that work: If p is 150 then m is 50, $\frac{1}{2} p$ is 75, and $\frac{1}{2} m$ miles is 25 miles. The total distance is (75 + 150 + 50 + 25) miles, or 300 miles. A diagram makes this problem an easy one to solve.)

6. Darla, $0.75; Ellie, $1.50; Fran, $3.75

7. a. Your drawing likely shows 2 boxes for Tom and 3 boxes for Ulysses. Tom is 2/5 of the bar, Ulysses 3/5.

    b. Your drawing might show 3 boxes for Vicky, 4 boxes for Willie, and 8 boxes for Xavier. Vicky ate 1/5 of the bar, Willie ate 4/15 of the bar, and Xavier ate 8/15 of the bar.

    c. Your drawing might show 1/3 of the bar for Yolanda, leaving 2/3 of the bar, so Zeb ate $\frac{3}{4} \times \frac{2}{3} = \frac{1}{2}$ of the bar, leaving $1 - (\frac{1}{3} + \frac{1}{2}) = \frac{1}{6}$ of the bar for Arnie.

8. a. Tom, $1.20; Ulysses, $1.80

    b. Vicky, $0.60; Willie, $0.80; Xavier, $1.60

    c. Yolanda, $1.00; Zeb, $1.50; Arnie, $0.50

9. 81 M&M's. After Tuesday, there were 5/6 of the original M&M's remaining. After Wednesday, there were $\frac{2}{3} \times \frac{5}{6} = \frac{5}{9}$ of the original M&M's remaining. If 5/9 of the original

number is 45, then 1/9 of the original number M&M's is 1/5 of 45, or 9, and then 9 one-ninths of the M&M's would be 81.

Answers for Chapter 9: Ratios, Rates, Proportions, and Percents

Supplementary Learning Exercises for Section 9.1

1. Perhaps. "Let's look at drawings. (Make rough drawings.) So the 105 by 100 does look more nearly square. The ratio of the two dimensions is more informative than the difference of the two dimensions."

2. All of the ratios, length:width, length:height, and width:height are closer to 1 for Box 2, so Box 2 is more nearly a cube.

3. In parts (b) and (d), a ratio would be appropriate (# lemons or salt:amount of water). And in part (a), it could be the ratio, maximum weight lifted to the weight of the lifter, or just the maximum weight the person could lift, and in part (c), it might be the amount of force that would bend the bar.

4. The ratio of the real and mirror heights of your image will be quite different from the ratio of the real and mirror widths of your image, so you will look distorted.

5. The results should taste different, with 5 times as much chili powder but only 1½ times as much garlic powder in the second version.

6. Although his daughter does have twice as many hits, she has batted more than twice as many times.

Supplementary Learning Exercises for Section 9.2

1. a. A proportion is an equality between two ratios.

    b. A rate is ratio of quantities that change but leave the value of the ratio the same.

2. a. To sing the same song would take the group of 3 singers about the same amount of time as it takes one singer.

    b. A larger crew should be able to do the job in less time, not more time.

3. a. $\frac{15}{16}$ mile (What was *your* thinking?)

    b. $8\frac{1}{8}$ miles   c. $2\frac{1}{2}$ inches   d. 24 inches   e. $\frac{1}{5}$ inch

4. a. $1\frac{1}{15}$ cups of sugar   b. $1\frac{19}{45}$ cups of sugar (Now you know why recipes are usually just halved or doubled, rather than adjusted to fit the amount of ingredients available!)

    c. $3\frac{3}{4}$ cups of flour   d. $11\frac{1}{4}$ cups of flour

    What methods did you use—mechanical proportion solving? Unit rates?

5. a. 2.4% loss in fuel efficiency   b. 2.4% gain in fuel efficiency
    c. 1.05% loss in fuel efficiency   d. $166\frac{2}{3}$ lbs reduction in weight
    e. 200 lbs increase in weight

6. a. Although the conclusion (that Bob has the best ratio) is correct, the *reasoning* does not take into account how many errors Bob took.

    b. The thinking does not take into account that Andy also had the fewest assists.

    c. Here are the assist to turnovers ratios: Andy, 4 to $10 = 0.4$ to 1, usually reported as 0.4, understood to be a unit ratio; Bob, 13 to $17 = \dfrac{\frac{13}{17}}{1} \approx \dfrac{0.76}{1}$, or just 0.76; and Cam,

    9 to $13 = \dfrac{\frac{9}{13}}{1} \approx \dfrac{0.69}{1}$. So Bob has the best ratio.

7. No. The team may have won 8 and lost 6, or won 12 and lost 9, for example. Ratios are often simplified.

8. a. $\dfrac{2\frac{2}{3}}{1}$ (or $2\frac{2}{3}$ to 1)   b. $\dfrac{\frac{3}{8}}{1}$ c. $\dfrac{2\frac{4}{5}}{1}$

9. The ratio of this year's budget to last year's budget has been 105% to 1 for the last ten years.

10. Dale's rate is $\dfrac{\$30}{2\frac{1}{2}\text{ h}} = \dfrac{\$12}{1\text{ h}}$, or \$12 per hour. So in 6 hours, Dale can expect to earn \$72.

11. a. $\dfrac{\text{number of births}}{92{,}228{,}000} = \dfrac{30.1}{1000} = \dfrac{92{,}228 \times 30.1}{92{,}228{,}000} \approx \dfrac{2{,}776{,}063}{92{,}228{,}000}$, so about 2,776,000 births.

    b. $\dfrac{4{,}058{,}814}{\text{pop in 1000s}} \approx \dfrac{14.7}{1000} = \dfrac{4{,}058{,}814 \div 14.7}{\text{pop in 1000s}} \approx \dfrac{276{,}110}{\text{pop in 1000s}}$, so about 276,110,000 people.
    (Notice that cross-multiplication *avoids* having to think about the situation even though it might lead to the same calculations!)

12. a. Half of 313,232,000 is 156,616,000, so over these females' lifetimes, the number of births could add about $156{,}616{,}000 \times 2.06 = 322{,}628{,}960$ children to the population.

    b. TRFs less than 2, if they do not increase, predict that a country's population will decrease (the two parents would not be replaced).

    c. A young woman—probably the TFR, because that gives an idea of the culture's typical number of children she will have. The public health official would probably find the birth rate more compelling in the short term, because it will suggest how the population will grow in the near future.

13. a.   The ratio, number of beats to time, stays the same for different intervals of time (or for different numbers of beats).

b.   Yes; 10 seconds is 1/6 of a minute, so 1/6 of 72 = 12 gives the number of beats. (Sometimes children think that a "per minute" rate does not apply to less than a minute.)

Supplementary Learning Exercises for Section 9.3

1. $\frac{697}{1000}$ is easily expressed in terms of hundreds ($\frac{69.7}{100}$, so = 69.7%), whereas $\frac{7}{13}$ will require calculation.

2.   a. 75% (100% of the original price − 25% of the original price)
     b. 60%   c. 85%   d. 50%   e. 75%   f. 100%

3.   a. ≈ 6.7%   b. $4\frac{1}{6}$%   c. $6\frac{1}{4}$%, $6\frac{1}{4}$%   d. $30,000
     e. About $3:$20, or 15% (approximating the increase of $2.95 by $3 and $19.95 by $20)
     f. 20,000 before; 18,800 after (Each 1% loss in the before employment = 200 people.)
     g. The *actual* change in the tax is 1%, but that 1% is a 10% increase over the old 10% tax.

4.   a. About 30%   b. About 25%   c. About $9   d. $2.49 (drawing?)
     e. About 40¢   f. $18.94

5.   a.   About 2738
     b.   About 58,460
     c.   Between the polling time and the election, events might happen that significantly influence voters, so the 37% could go up or down.

6.   a.   6.2 hot-dogs per minute
     b.   Continuing to eat hot-dogs at that rate for an hour is not reasonable.
     c.   One way: In 4 minutes, the champion would eat 24.8 hot-dogs, and the challenger would eat 21.2. So the challenger would be 3.6 hot-dogs behind. Another way: The difference in their rates, 6.2 − 5.3 = 0.9 hot-dogs per minute, gives a difference of 3.6 hot-dogs in 4 minutes.
     d.   Not likely. Most eaters likely slow down as they continue to eat. You could also argue that the *top* finishers, on the other hand, may be able to maintain a constant rate.

7.   a.   Proportion: $110:9 = x:14$, $x \approx 171$ calories. Unit rate: $110:9 = 12\frac{2}{9}:1$; so for 14 crackers, $14 \times 12\frac{2}{9} = 171\frac{1}{9}$ calories, or 171 calories. Similarly, about 5.4 grams of fat.

b.   Proportion way: $\dfrac{3.5}{9} = \dfrac{65}{y}$, $y \approx 167$ crackers

c. $\dfrac{3.5}{65} \approx 0.0538$, so about 5%.

Supplementary Learning Exercises for Chapters 7, 8, and 9

1. Lisa's rate was 1.5 ounces of white per drop of green. Since Rachel had the same shade, her rate of white to green must have been the same. Since Rachel used 6 drops of green, she must have used $6 \times 1.5 = 9$ ounces of white. Alternatively, solve $\frac{6}{4} = \frac{x}{6}$.

2. Drawings may be helpful here. With $\frac{3}{4}$ gallon, Todd should be able to paint $\frac{3}{4}$ as much as he could with 1 whole gallon, or $\frac{3}{4} \times \frac{3}{5} = \frac{9}{20}$ of the wall. For the entire wall question, since 1 gallon paints $\frac{3}{5}$ of the wall, $\frac{1}{3}$ gallon would paint $\frac{1}{5}$ of the wall, and it would take 5 times as much to paint the whole $\frac{5}{5}$ of the wall, or $5 \times \frac{1}{3} = \frac{5}{3} = 1\frac{2}{3}$ gallons of paint. The two questions could be addressed "mechanically" with $\frac{\frac{3}{5}}{1} = \frac{x}{\frac{3}{4}}$ and $\frac{\frac{3}{5}}{1} = \frac{1}{x}$.

3. In 8 hours, the worker whose rate is 36 parts per hour will make 288 parts. That means the other worker made $512 - 288 = 224$ parts in the 8 hours, a rate of 28 parts per hour.

4. A drawing may be helpful here. The 16 liters fill $\frac{2}{5}$ of the tank, so 8 liters would fill $\frac{1}{5}$ of the tank, and $5 \times 8 = 40$ liters would fill the whole tank. Its volume is 40 liters. Can you also use a proportion here?

5. This situation is just repeated addition, so John needs $15 \times \frac{3}{8} = \frac{45}{8} = 5\frac{5}{8}$ meters of ribbon. Notice that the proportion $\frac{\frac{3}{8}}{1} = \frac{x}{15}$ could be used. The (advanced) proportional reasoning has repeated addition as a special case.

6. Since 6 trees represent $\frac{3}{2}$ of the trees, 2 trees would be $\frac{1}{2}$ of the trees, so there would be 4 trees in the park. Alternatively, $\frac{6}{\frac{3}{2}} = \frac{x}{1}$, where the unit for the $\frac{3}{2}$ and the 1 is the number of trees in the park.

7. If $\frac{1}{3}$ of the cars has 18 cars, then all $\frac{3}{3}$ of the cars would have 3 times as many, or 54.

8. Similarly, if the 9 dogs represent $\frac{3}{7}$ of all the dogs, then 3 dogs would represent $\frac{1}{7}$ of the dogs. So the whole $\frac{7}{7}$ of the dogs at the pound would be 7×3, or 21, dogs.

9. One way, focusing on 7 days per week, 3.5 kg per week per 10 persons: 50 people, then, would take $5 \times 3.5 = 17.5$ kg per week. In 28 days (4 weeks), they would require $4 \times 17.5 = 70$ kg of sugar.

10. Ten small glasses will fill $\frac{10}{15}$ or $\frac{2}{3}$ of the large jar, or what nine large glasses will fill. So $\frac{2}{3} \times 9 = 6$ large glasses will fill the small jar. Alternatively, one small glass will fill $\frac{9}{15}$ of a

large glass, so 10 small glasses would match $10 \times \frac{9}{15}$ or six large glasses. Still another alternative: Solve $\frac{15}{9} = \frac{10}{x}$. Can you explain where this proportion comes from?

11. In one week, the horse will eat $10 \div 3 = 3\frac{1}{3}$ pounds of hay, so in five weeks, the horse will eat $5 \times 3\frac{1}{3} = 16\frac{2}{3}$ pounds of hay.

12. The 15 chips represent $\frac{5}{6}$ of the unit, so $\frac{1}{6}$ of the unit would be three chips, and $6 \times 3 = 18$ would represent the whole $\frac{6}{6}$ of the unit. Then $\frac{2}{3}$ of the unit would be 12 chips, and $1\frac{1}{2}$ of the units would be 27 chips. The questions could also be handled by proportions; can you write them?

13. 40 kilometers in 16 minutes is a rate of 2.5 kilometers per minute. So in 36 minutes, the trip should cover $36 \times 2.5 = 90$ kilometers. Alternatively, 40 minutes is 2.5 times as much as 16 minutes, so the train should cover 2.5 times as much in 40 minutes as it does in 16 minutes: $2.5 \times 36 = 90$ kilometers. Another method would use a proportion: $\frac{16}{40} = \frac{36}{x}$.

14. Beeburg. Antville's budget was \$12M, as was Cowtown's, but Beeburg's was \$12.5M.

15. 100,000. The unit for the five-eighths is Middletown's population, suggesting looking at eighths of M's population. Then M's excess population of 15,000 over L's shows that each eighth of M's population is 5000. L then has 25,000 and B 100,000.

16. 48. A drawing might suggest an analysis leading to the following reasoning. The known fractions eaten give $\frac{1}{8} + \frac{1}{8} + \frac{1}{8} + \frac{1}{3} + \frac{1}{4}$, or $\frac{23}{24}$ of the bag. So Fran's two chocolates were $\frac{1}{24}$ of the bag, and the whole bag ($\frac{24}{24}$) contained 48 chocolates.

17. Because they paint at the same rate, this amounts to an additive situation. Joel is 80 square yards ahead of Kendrick when Kendrick starts, so he will be 80 square yards ahead when he has painted 600 square yards. Kendrick will have painted 520 square yards.

18. This situation is proportional, with the Lewis to Max ratio of square yards being 120:40 or 3:1. So, Max will have painted 200 square yards when Lewis has painted 600.

19. Tyler: \$7 per hour; Ullie: \$8.50 per hour.

20. Velma will pay $\frac{1}{3}$ of the rent, Winnie will pay $\frac{4}{15}$, and Zoe will pay $\frac{2}{5}$.

Answers for Chapter 10: Integers and Other Number Systems

Supplementary Learning Exercises for Section 10.1

1. From smallest to largest, $^-\left(\frac{53}{6}\right)$ $^-\left(\frac{53}{7}\right)$ $^-\left(\frac{5}{4}\right)$ $^-\left(\frac{3}{4}\right)$ $^-\left(\frac{1}{10,000}\right)$ $\frac{1}{1000}$ $\frac{53}{7}$ $\frac{53}{6}$

2. a. 2 or +2   b. –3   c. –3   d. 10 or +10

Supplementary Learning Exercises for Section 10.3

1. a. 0   b. $^-$2   c. $^-$13

2. 5 chips for positive, 3 for negative. The student is merely counting chips and not evaluating according to the "canceling" effect of two chips of opposite colors.

3. A gain of 4 yards, a gain of 8 yards, losses of 3 and 15 yards, a gain of 11 yards

4. If the child accepts the view of, say, 2 and $^-$2 as being on the opposite sides of 0, then it may be plausible that $^-$0 must belong to the same point as 0.

Supplementary Learning Exercises for Section 10.4

1. a. Five chips of one color for 5 and 2 chips of an "opposite" color for $^-$2, giving 3 chips of the color for positive after pairs of opposite colors cancel.

    b. (one way)

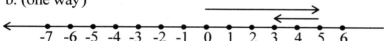

    c. Sample: In the mail, you received a rebate of $5 but an overdue charge of $2 for a bill you sent in late. How has your financial position changed?

2. a. Show 5 with 7 positive chips and 2 negative ones. Removing two negative chips gives 7.
    b. Show 5 negative chips and take away 2 of them. The remainder is $^-$3.
    c. Show $^-$5 with 7 negative chips and 2 positive chips. Take away the two positive chips, leaving $^-$7 as the answer.

3. a. $^-$78   b. $^-$468   c. $^-\left(\frac{17}{24}\right)$ d. $\frac{1}{2}$ e. 1.713

4. a. 57,000 + 35,000 + (16,000) + (16,000)
    b. $750.80   ($1000)   ($110.50)

5. All of your choices should suggest that $^-(a-b) = \, ^-a+b$.

Supplementary Learning Exercises for Section 10.5

1. Start with, say, $4 \times {}^-5 = {}^-20$, $3 \times {}^-5 = {}^-15$, $2 \times {}^-5 = {}^-10$, etc., continuing to ${}^-2 \times {}^-5$, and look for a pattern.

2. a 1088.7   b. $\frac{1}{20}$ c. ${}^-128$ d. 413.6

3. With simple calculators, the user must assign the sign to the product.

4. a. and b. Show the dividend (the ${}^-10$ and the ${}^-16$) and repeatedly remove chips for the divisor (–2). The number of groups removed will be the quotient.

5. Compare your favorite with those of others.

6. a. Possibilities: "The additive inverse of the product of *a* and *b*;" "the additive inverse of *a*, times *b*;" and "*a* times the additive inverse of *b*."

   b. All of your choices should suggest that the three expressions are all equal.

Supplementary Learning Exercises for Section 10.6

1. There is no whole number between consecutive whole numbers, like 8 and 9.

2. a. Using equivalent fractions ${}^-(\frac{300}{400})$ and ${}^-(\frac{100}{400})$, it is easy to see several.
   b. Infinitely many

3. a. Possibilities, all keeping the order of the factors the same: $\frac{5}{8} \times [\,{}^-20 \times ({}^-4\frac{5}{6} \times \frac{8}{35})]$; or
   $[(\frac{5}{8} \times {}^-20) \times {}^-4\frac{5}{6}] \times \frac{8}{35}$; or further applications to the expressions in square brackets.
   b. Change the order of the two factors within either set of parentheses, or change the order of the parenthetical expressions, $({}^-4\frac{5}{6} \times \frac{8}{35}) \times (\frac{5}{8} \times {}^-20)$.
   c. $(\frac{5}{6} \times 24) + (\frac{5}{6} \times {}^-600)$ d. $\frac{2}{3} \times ({}^-298 + {}^-2)$
   e. $(2 + {}^-5)x$ f. 1×1 g. $5 \times \frac{1}{5}$

4. a. Samples: $({}^-15 + 8) + (7 + {}^-3)$; or $({}^-3 + 7) + ({}^-15 + 8)$; or $(7 + {}^-3) + (8 + {}^-15)$; or combinations of those three
   b. $0 + 0$
   c. $5 + {}^-5$
   d. Samples: $[({}^-2 + {}^-4) + 5] + {}^-5$; or ${}^-2 + [({}^-4 + 5) + {}^-5]$; or further uses of associativity on the expressions in square brackets. Notice that the order of the addends does not change.

5. a. Yes, the set, –1, 0, 1, is closed under multiplication because the product of every choice of pairs of numbers from the set is also a number in the set.
   b. The set, –1, 0, 1, is not closed under addition because, for example, $1 + 1 = 2$, and 2 is not in the set.
   c. The set of numbers, –2, 0, 2, is not closed under multiplication because, for example, $2 \times 2 = 4$, and 4 is not in the set.

Answers for Chapter 11: Number Theory

Supplementary Learning Exercises for Section 11.1

1. A prime number is a whole number that has exactly two different factors. A composite number is a whole number greater than 1 that has more than two factors.

2. No, composite numbers by definition have more than two factors, and prime numbers are restricted to numbers with exactly two factors.

3. True. A prime number p can be expressed only by $p \times 1$ or $1 \times p$.

4. True. Each number crossed out is a multiple of some number greater than 1, so that number would be a third factor of the number crossed out.

5. b. 361. 23 won't be crossed out (it is the next prime), and 323 is already crossed out, when all the multiples of 17 were crossed out. 20 is also already crossed out, as a multiple of 2.

6. No. For example, 2 is a factor of 6, but 6 is not a factor of 2.

7. Many examples exist. For $2 \times 4 = 8$, for example, "8 is a multiple of 2" and "2 is a factor of 8" are both correct.

8. a. 87 is a factor or 2088.   b. 24 is a factor of 2088.

    c. 2088 is a multiple of 87, or 2088 is a multiple of 24.

9. a. g is a factor of w.   b. s is a factor of w.

    c. w is a multiple of g, or w is a multiple of s.

10. a. 1, 7, and 49 are the only factors of 49 because 49 is the square of 7.

    b. Yes, 49 is a composite number because it is larger than 1 and has more than two factors.

    c. 0, 49, 98, 147, 196, 245…. Multiply 49 by any whole number to get a multiple of 49.

11. a. 31 is a prime because it has only 1 and 31 as factors.

    b. 299 is a composite because 13 (or 23) is a third factor besides 1 and 299.

    c. 27 is a composite because 3 (or 9) is a third factor besides 1 and 27.

    d. 999 is a composite because 3 (and 333 or 111 or …) would be a third factor besides 1 and 999.

    e. "Prime" and "composite" refer only to whole numbers, not fractions different from whole numbers.

12. No, because the product of two primes would have more than two factors and thus not be a prime.

13. a. 0 is the only multiple of 0.

    b. Every whole number n is a factor of 0 because $n \times 0 = 0$.

14. Adding the prime number 2 to any other prime number will give a counterexample.

15. Not only 1 and 9991 are factors of 9991. 97 (and 103) are also factors, so 9991 has more than two factors.

16. If a number is composite, it will be a multiple of its smallest factor (not 1) and so would have been marked out when multiples of that factor were marked out.

17. 4 (1, 2, 4) is another one, as are 25 (1, 5, 25) and 121 (1, 11, 121), along with the given 49 (1, 7, 49) (*Hint*: $25 = 5^2$)

Supplementary Learning Exercises for Section 11.2

1. No. 2^{10} is just shorthand for $2 \times 2 \times 2 \times 2 \times 2 \times 2 \times 2 \times 2 \times 2 \times 2$, which is a prime factorization.

2. The UFT says that *every* prime factorization of 1024 will involve exactly ten 2s as factors.

3. There are different (correct) factor trees for each part, except part (e), which as a prime can have, at most, branches to 29 and 1, and usually branches that end in 1 are not put in. Whatever your factor tree, you should have the following prime factorizations (with your factors possibly in a different order, of course).

    a. $960 = 2^6 \times 3 \times 5$ b. $9600 = 2^7 \times 3 \times 5^2$ c. $1125 = 3^2 \times 5^3$

    d. $8100 = 2^2 \times 3^4 \times 5^2$

4. No, these different starts do not violate the Fundamental Theorem of Arithmetic because the theorem applies only to prime factorizations, and neither Kim nor Lee has a prime factorization yet.

5. If k is an odd number, it cannot have 2 as a factor, which is implied by $k = 2m$.

6. Neither Abbie nor Bonita has a complete prime factorization. $57 = 3 \times 19$, and $171 = 3^2 \times 19$, so with that further work, the complete prime factorizations would agree.

7. a. 2, 3, 11, and 19 are the only prime factors of 3,972,672.

    b. There are 164 possibilities! Some, besides 33, are 121, 57, 6, 12, 24, 209,… Just be certain that your composite factors involve only a selection of the prime factors in the prime factorization.

8.   No. Unique factorization into primes assures that 7^{15} has only 7 as a prime factor, and that 9^m has only 3 as a prime factor. Thus, the two can never be equal.

9.   a. 630 (Did you realize that this is just the LCM?)

     b. 990   c. 2940

10. a. True. A nonzero multiple of 75 will be $75n$ for some whole number n greater than 1, so the 3×5^2 must appear in the prime factorization of $75n$, by the Unique Factorization Theorem.

     b. True, with reasoning like that in part (a).

     c. The smallest such number must be $2^3 \times 3 \times 5^2 = 600$.

11. Both sides involve only prime factors of 5, so it may be possible. $(5^2)^m = 5^{2m}$, and $(5^3)^n = 5^{3n}$, so any values of m and n that make $2m = 3n$ will be possible values. One example is $m = 3$ and $n = 2$. Any (nonzero) multiples of those will also work.

12. a. $(2 + 1)(3 + 1)(1 + 1) = 3(4)(2) = 24$ factors

     b. $(x + 1)(y + 1)(z + 1)$ factors

13. One million $= 1,000,000 = 10^6 = (2 \times 5)^6 = 2^6 \times 5^6$, so one million has $7 \times 7 = 49$ factors.

14. 100, reasoning as in Exercise 12. Two are primes: 2 and 5. 1 is a factor. So the other 97 factors must be composites.

15. At least one 3 and at least one 7 must appear in the prime factorization of the number. Each could be raised to powers so that, from $15 = 3 \times 5$, one exponent is 2 and the other 4: $3^2 \times 7^4$, or $3^4 \times 7^2$, for example.

16. Yes, it is possible. For example, start one factor tree for 36 with 6×6 and a second factor tree for 36 with 4×9. The two factor trees will be different when completed. (But the prime factorization obtained through the different factor trees are the same.)

Supplementary Learning Exercises for Section 11.3

1.   Relatively prime numbers have only 1 as a common factor. So the pairs in parts (a), (c), (d), and (e) are relatively prime. In part (b), the numbers have 5 as a common factor.

2.   True

3.   a. 2, 3, 4, 6, and 9 are divisors.   b. 2, 4, 5, 8, and 10 are divisors

     c. 2, 3, 4, and 6 are divisors.   d. 3 and 5 are divisors.

     e. None of 2, 3, 4, 5, 6, 8, 9, 10 is a factor.   f. Only 2 is a divisor.

g. Each of 2, 3, 4, 5, 6, 8, 9, and 10. (Was it necessary to multiply 80×54 out?)

h. 2, 4, 5, 8, and 10 are divisors.

4.   With approximate square roots as the guide…

   a. 19 ($\sqrt{401} \approx 20$ and we need check only primes)

   b. 19 (23^2 would be too large)   c. 29   d. 37

5.   a. $207 = 3^2 \times 23$ b. $121 = 11^2$ c. 83×71 d. 19^3

   e. $247 = 13 \times 19$ f. $119 = 7 \times 17$ g. Prime   h. Prime   i. $297 = 3^3 \times 11$

6.   Check that the sum of the digits in your number has 3 as a factor, but not 9, that the number named by the rightmost two digits has 4 as a factor, and that the number named by the rightmost three digits does not have 8 as a factor.

7.   If 9 is a factor, then 3 is automatically a factor (because $9 = 3 \times 3$).

8.   It is easy to see that 3 and 9 are factors of $(999 + 5 \cdot 99 + 2 \cdot 9)$.

Supplementary Learning Exercises for Section 11.4

1.   a. 12   b. 144   c. Eliminate the GCF: $\frac{12 \times 3}{12 \times 4} = \frac{3}{4}$ d. $\frac{17}{36} + \frac{5}{48} = \frac{68}{144} + \frac{15}{144} = \frac{83}{144}$

2.   a. 30   b. 1350   c. $\frac{150}{270} = \frac{30 \times 5}{30 \times 9} = \frac{5}{9}$ d. $\frac{91}{150} + \frac{7}{270} = \frac{819}{1350} + \frac{35}{1350} = \frac{854}{1350} (= \frac{427}{675})$

3.   a. LCM = 1056; GCF = 12

   b. LCM = 126; GCF = 42 (Did you see this one fast?)

   c. LCM = 576; GCF = 4

   d. LCM = $x^4 y^9$; GCF = $x^2 y^6$

   e. LCM = $2^3 \times 7^3 \times 11 \times 13 = 392{,}392$; GCF = $2^2 \times 7^2 = 196$

4.   a. 1800

   b. The algorithm involves finding a common factor, then "removing" it by dividing, and continuing, removing other common factors. At the end, then, the common factors multiplied are the least ones whose product will be a common multiple of the original numbers.

5.   a-b. Yes, at any common multiple of 2 and 3: 6, 12, 18, 24, etc.

Part II: Reasoning About Algebra and Change

Answers for Chapter 12: What Is Algebra?
Supplementary Learning Exercises for Section 12.1

1. a. Distributive property: $1(x + 3) + 5(x + 3) = 6(x + 3)$

    b. Distributive property (plus computations)

    c. Associative property of addition

    d. False

    e. Rewriting 16 and then using the associative property of multiplication

    f. Distributive property (then computation)

    g. False

    h. False

    i. Distributive property

    j. Associative property of multiplication

    k. False

    l. Zero property of multiplication

    m. Commutative property of addition

    n. Identity property of multiplication

2. a. For any value b: $b = b$ is always true; $b \leq b$ is always true; $b \geq b$ is always true.

    b. For any values m and n: If $m = n$, then $n = m$ is always true; if $m \leq n$, then $n \leq m$ is sometimes true (but only when $m = n$); if $m \geq n$, then $n \geq m$ is sometimes true (but only when $m = n$).

    c. For any values r, s, and t: If $r = s$ and $s = t$, then $r = t$ is always true; if $r \leq s$ and $s \leq t$, then $r \leq t$ is always true; if $r \geq s$ and $s \geq t$, then $r \geq t$ is always true.

3. a. True

    b. False. This formula applies only to squares.

    c. True. If each side of the square is $2s$, then the area is $(2s)^2$, or $2^2 s^2$, or $4s^2$.

d. True. The perimeter of any polygon is the sum of the sides.

e. False. The circumference is 10π. (You may not remember the formula, $C = 2\pi r$, but you should remember, at least, that π is in the formula.)

4. Let c represent the number of comic books owned by Casey; k represent the number of comic books owned by Kaye; n represent the number of comic books owned by Nancy; j represent the number of comic books owned by Jinfa; f represent the number of comic books owned by Fortuna; and m represent the number of comic books owned by Milissa.

a. $k = 3c$ b. $n > 2c$ c. $j = \frac{1}{2}c$ d. $f = 3c + 4$ e. $m = 3c + 2$

5. a. $n + (n + 2) + 2n = 4n + 2$ b. $(n + 2) + (n + 2) + (2n - 6) = 4n - 2$ c. $4n - 8$

6. a. Yes   b. No   c. Yes   d. No   e. Yes   f. No

7. a. $120 million + $z = $560 million$, or more simply, $120 + z = 560$, so $z = 560 - 120 = 440$. Freddie's company's mortgage debt is \$440 million.

b. Let t represent the degrees of temperature lost: $104° - t = 99.6°$. So $t = 104° - 99.6° = 4.4°$.

c. Let a represent the entrance fee in dollars for Air Park for one person. Then $5a$ represents the dollar cost for 5 people. Thus, $5a + 5 \cdot 32 = 300$, so $5a = 140$ and $a = 140 \div 5 = 28$ dollars. The entrance fee for Air Park was \$28.

d. Let d equal the price of one day of the three-day pass. Then $3d = \$81$, so $d = \$27$. Tanya should buy the three-day pass and save \$5 per day, or \$15 for the 3 days.

e. The price p of 1 ounce of nuts can be used to connect the fact that there are 16 ounces in 1 pound, and the cost of 1 pound $16p = 3.68$, so $p = \$0.23$. Six ounces would cost $6 \times \$0.23 = \$1.38$.

f. Let j represent the cost per ounce of the jellybeans. $8 \times \$0.23 + 8j = \$3.28$, so $j = \$0.18$.

8. a. 5900, using associativity of multiplication

b. \$39.80, using distributivity: $(4 \times 3.98) + (6 \times 3.98) = (4 + 6) \times 3.98 = 10 \times 3.98$

c. $\frac{97}{106}$, using multiplicative inverses and the identity for multiplication

d. $144, using distributivity: $(24 \times 1.50) + (24 \times 4.50) = 24 \times (1.50 + 4.50) = 24 \times 6$

9.   a. $b + 7$ b. $2b + 9$ c. $b + c$ d. $2b + 12$

10. a. $x = 4$ b. $x = 2.4$ c. $^-28$ d. $^-10$ e. 547

11. a. The result is not 25. "Add 11" should be "Add 13."
     b. Suppose your age is a. Write each of the steps using a instead of your age, then simplify the expression you have. The result is 25. To obtain your age as the result, subtract 25 instead of subtracting your age.

Supplementary Learning Exercises for Section 12.2

1. a. $4x^4 + x^3 + 2x + 5$ (It is okay to include $0x^2$.)

    b. $2x^5 + 3x^4 + 4x^3 + x^2 + 2x + 5$

2. a. $9x^2 + 6x + 5$

    b. $18x^2 + 8x + 6$

    c. $4 + 3x + 8x + 6x^2$, or $6x^2 + 11x + 4$

3   a.   Sum: $12x^2 + 5x + 19$; product: $35x^4 + 31x^3 + 119x^2 + 48x + 90$

    b.   Sum: $\frac{13}{12}x + 2\frac{2}{5}$, or $1\frac{1}{12}x + 2\frac{2}{5}$; product: $\frac{1}{4}x^2 + \frac{29}{30}x + \frac{4}{5}$

    c.   Sum: $8x + {}^-5$; product: $15x^2 + {}^-19x + 6$

    d.   Sum: $41x + 56$; product: $418x^2 + 1139x + 775$

    e.   Sum: $14x + 3$; product: $42x^3 + 127x^2 + 2x + {}^-3$

4. a. $\frac{3}{x}$

    b. $\frac{11}{2x}$

    c. $\frac{7-3y}{2(y+7)^2}$

    d. $\frac{2}{y^2}$

    e. $\frac{n}{z}$

Supplementary Learning Exercises for Section 12.3

1. a.   (i) This is an arithmetic sequence with $a = 8$ and $d = 4$. That is, the numbers increase by 4 each time, so by the 500th number, there will have been 499 increases of 4, or 1996, over the starting value of 8: $8 + 499(4) = 2004$.

(ii) The same thinking (or the use of the result for an arithmetic sequence) gives $8 + (n{-}1)4$ for the nth number. $8 + (n-1)4$ also $= 8 + 4n - 4 = 4 + 4n$.

b. This is an arithmetic sequence with $a = 2.3$ and $d = 1.3$.

  (i) $2.3 + 499(1.3) = 651$

  (ii) $2.3 + (n-1)1.3$, or $1.3n + 1$

c. (i) Every odd term is $^-1$ and every even term is 1, so the 500th number would be 1.

  (ii) The nth term would be $^-1$ if n is odd, and 1 if n is even.

d. This is an arithmetic sequence with $a = 3\frac{1}{4}$ and $d = 1\frac{3}{4}$.

  (i) $3\frac{1}{4} + 499(1\frac{3}{4}) = 3\frac{1}{4} + 873\frac{1}{4} = 876\frac{1}{2}$ (ii) $3\frac{1}{4} + (n{-}1)1\frac{3}{4}$, or $1\frac{3}{4}n + 1\frac{1}{2}$

e. This is an arithmetic sequence with $a = {}^-100$ and $d = 2$.

  (i) $^-100 + 499 \times 2 = {}^-100 + 998 = 898$ (ii) $^-100 + (n-1)2 = {}^-102 + 2n$

f. (i) The 4-7-9 pattern is a block of three digits. From $500 \div 3 = 166$ R 2, there would be 166 full blocks and two numbers into the next block, or 7.

  (ii) Calculate $n \div 3$ and use the remainder to see which number in the next 4-7-9 block to choose; if the remainder is zero, then there are full 4-7-9 blocks, so the last number would be 9.

g. This is a geometric sequence with $a = 1$ and $r = 5$.

  (i) The 500th number would be 5^{499}. (ii) The nth term is 5^{n-1}.

h. This is an arithmetic sequence with $a = 2000$ and $d = {}^-1.7$.

  (i) The 500th number would be $2000 + 499(^-1.7) = 1151.7$.

  (ii) $2000 + (n-1)(^-1.7)$, or $2000 - (n-1)1.7$, or $2001.7 - 1.7n$

i. This is a geometric sequence with a and r both equal to 3.

  (i) The 500th number would be $3(3^{500-1}) = 3^{500}$.

  (ii) The nth term would be $3(3^{n-1}) = 3^n$.

j. This is a geometric sequence with $a = 1$ and $r = \frac{1}{2}$.

  (i) The 500th number would be $1 \cdot (\frac{1}{2})^{500-1} = \frac{1}{2^{499}}$.

  (ii) The nth term would be $\frac{1}{2^{n-1}}$.

k. This is an arithmetic sequence with $a = 16$ and $d = {}^-4$.

   (i) The 500th term would be $16 + 499({}^-4) = {}^-1980$.

   (ii) The nth term would be $16 + (n - 1)({}^-4) = 20 - 4n$.

l. This is an arithmetic sequence with $a = 0.5$ and $d = 0.25$.

   (i) The 500th term would be $0.5 + 499(0.25) = 125.25$.

   (ii) The nth term would be $0.5 + (n - 1)(0.25)$, or $0.25n + 0.25$.

2. a. 22.3, 22.65, 23, 23.35, 23.7   b. 1.5, 3, 6, 12, 24

3. Here is one way of reasoning (cf. 1(f) for another way). Notice that the 1st, 4th, 7th etc. digit is 3. The 2nd, 5th, 8th etc. digit is 6. The 3rd, 6th, and 9th digit is 9. The 36th digit is in the sequence 3, 6, 9,... so the digit is 9. The 100th digit is in the sequence 1, 4, 7, ... so the digit is 3. (Notice that 100 is one more than 99, and the digit in the 99th place would be 3 because 99 is in the sequence 3, 6, 9, ... The number following 9 in the given pattern of 369369369...is 3.)

4. $\frac{2}{11} = 0.18181818...$. Notice that the even-numbered digits (2nd, 4th, etc.) are all 8s and the odd numbered digits are all 1s. The 88th digit would therefore be 8.

5. The calculations 2, $2^2 = 4$, $2^3 = 8$, $2^4 = 16$, $2^5 = 32$, $2^6 = 64$, $2^7 = 128$, $2^8 = 256$, $2^9 = ...2$, $2^{10} = ...4$, ...8, ...6, ...2, ... suggests that the 2-4-8-6 block of four digits is repeated. From $120 \div 4 = 30$, you know that there will be 30 full blocks of 2-4-8-6 when you get to 2^{120}, so the last digit, 6, will be in the ones place.

6. 85^2 would end in 25. It would begin with 8×9, or 72. Thus, $85^2 = 7225$. Similarly, $19.5^2 = 380.25$, from $19 \times 20 = 380$. (195^2 would equal 38,025.)

7. a. 9   b. 7   c. 1

8. a. $\dfrac{1}{3^{n-1}}$ (or perhaps $3^{-(n-1)} = 3^{-n+1}$)   *Hint*: Make a table:

Term	Number in sequence	Number in sequence rewritten with term number
1	$1 \quad = 3^0$	3^{-1+1}
2	$\dfrac{1}{3} = \dfrac{1}{3^1}$ or 3^{-1}	3^{-2+1}
3	$\dfrac{1}{9} = \dfrac{1}{3^2}$ or 3^{-2}	3^{-3+1}
4	$\dfrac{1}{27} = \dfrac{1}{3^3}$ or 3^{-3}	$3^{?}$
.	.	.
.	.	.
.	.	.
n	$\dfrac{1}{3^?}$	$3^{?}$

b. The sum of the set of numbers *after* the first term (1) will get closer and closer to $\frac{1}{2}$, so 1 (the first term) plus the sum of the set of numbers in the remainder of the sequence will get closer and closer to $1\frac{1}{2}$.

Supplementary Learning Exercises for Section 12.4

1. a. Function: Every passenger is matched with exactly one seat.

b. Not a function: Not every seat is always matched with a passenger.

c. Function: A, B, and C in the first set are each matched with exactly one element in the second set.

d. Not a function: One element for the first set, namely A, is matched with not one but three elements in the second set.

e. Not a function: 4 can be matched with both 2 and ‾2.

f. Function: Every number has exactly one square.

g. Function: Every number (the first in each ordered pair) is matched with exactly one number.

h. Not a function: For example, 3 (the first number in an ordered pair) is matched with both 6 and 8.

2. a. $y = 5x + 7$, $m = 507$, $n = 400$

   b. $f(x) = 157 - 4x$, $m = 13$, $n = 50$

   c. $output = 2^{input} + 3$, $m = 1027$, $n = 11$

   d. $output = 3 \times input + 5$; $m = 305$, $n = 212$

   e. $g(x) = x^3 + 1$, $m = 1001$, $n = 20$

   f. Notice that the lines for the x-$h(x)$ values need to be put in order. $h(x) = 7x - 3$, $m = 172$, $n = 351$.

3. a. $f(n) = 7n + 1$. One justification: In the nth shape, going sideways one "layer" at a time, there are $n + 2n + (n+1) + 2n + n$, or $7n + 1$, toothpicks required.

   b. $f(n) = 3n + 8$. One justification: In the nth shape, across the top takes $n + 1$ toothpicks; the vertical pieces take $n + 2$; and the bottoms of the squares take another $n + 1$. The slanted sides of the triangles take 4 toothpicks. The total is $(n + 1) + (n + 2) + (n + 1) + 4 = 3n + 8$ toothpicks.

   c. $f(n) = 8n + 1$. One justification: In the nth shape, the bottoms of the hexagons take $2n$ toothpicks, the vertical segments take $n + 1$ toothpicks, the tops of the hexagons take $2n$ more, and each of the rest of the n squares requires 3, or $3n$ in all. The grand total is $2n + (n + 1) + 2n + 3n = 8n + 1$ toothpicks.

4.   The two types are alike in that each entry in the sequence, after the first one, is related to the previous one in the same way. For an arithmetic sequence, the relationship is adding (or subtracting) the same number each time, but for a geometric sequence, the relationship is multiplying by the same number each time.

5. a.   12, 39, 66, 93, 120, 147, 174

   b.   3, $9\frac{3}{4}$, $16\frac{1}{2}$, $23\frac{1}{4}$, 30, $36\frac{3}{4}$, $43\frac{1}{2}$

   c.   3, 15, 75, 375, 1875, 9375, 46875

   d.   8, $5\frac{1}{3}$, $3\frac{5}{9}$, $2\frac{10}{27}$, $1\frac{47}{81}$, $1\frac{13}{243}$, $\frac{512}{729}$ (Why do the entries keep getting smaller?)

6. 200 chirps; $C = 4T - 160$ (Below what temperature is the cricket not chirping?)

Supplementary Learning Exercises for Section 12.5

A drawing is usually a good idea. Be certain that you write clearly what any variable represents, including its unit. For each problem, the answer is given first and then an example of an equation that could be written. Other ways of thinking might lead to different equations, but the answer to the question should be the same. It is a good idea to check your solution by seeing whether it fits the story statement. If you miss a problem, was it because of your algebraic expressions, your algebraic equation, or your solving of the equation?

1. a. Each allowance was $8 a week. Let a = the amount of the weekly allowance ($). Then $a - 5 = \frac{1}{2}(a - 2)$.

    b. Each paid $2800 for his car originally. Let p = the initial cost of each car ($). Then $p + 200 = \frac{5}{6}(p + 800)$.

    c. Each outfit cost $120 originally. Let x = the original cost of each outfit ($). Then $x - 45 = \frac{3}{4}(x - 20)$.

2. a. First runner, 58.5 s; second runner, 60.5 s; third runner, 63.5 s; and fourth runner, 56.5 s (The information about the second runner refers to the first runner, so let f = the number of seconds for the first runner; then the other runners' times are $f + 2$, $(f + 2) + 3$, and $(f + 2) - 4$. Because 3:59 = 239 seconds, we have
    $f + (f + 2) + [(f + 2) + 3] + [(f + 2) - 4] = 239$ (you may have simplified early, getting $f + (f + 2) + (f + 5) + (f - 2) = 239$.)

    b. First swimmer, 102 s (or 1:42 min), and second swimmer, 104.5 s (or 1:44.5 min). Let f = the time for the first swimmer (seconds). Then $f + (f + 2.5) + [f + (f + 2.5)] = 413$.

    c. First run, 50.87 s, and second run, 49.81 s. Let x = time for the first run in seconds. Then $x + (x - 1.06) = 100.68$ because 1:40.68 minutes = 100.68 seconds.

3. a. Danetta $75, Elaine $50, and Jan $40. Let d = amount Danetta spent ($). Then $d + (d - 25) + (d - 35) = 165$. Or, if e = amount Elaine spent ($), then $e + (e + 25) + [(e + 25) - 35] = 165$.

    b. Danetta $18.50, Elaine $37, and Jan $30.50. Let J = amount Jan spent ($). Then $(J - 12) + 2(J - 12) + J = 86$.

4. a. First store $125.85, second $25.20, and third $16.75. Let t = amount spent at the third store ($). Then $t + (t + 8.45) + 3(2t + 8.45) = 167.80$.

    b. First store $14.15, second $0, third $30.10, and fourth $5.40. Let x = amount spent at the first store ($). Then $x + 0 + (x + 15.95) + (x - 8.75) = 49.65$.

5. a. Alima $\frac{1}{4}$ mile, Noor $\frac{3}{4}$ mile, Noor's husband $1\frac{1}{2}$ miles. Let N = the number of miles Noor carried the baby. Then $\frac{1}{3}N + N + 2N = 2\frac{1}{2}$.

b. Dien 200 min (3 h, 20 min); Gia 105 min (1 h, 45 min); Minh 210 min (3 h, 30 min). Let G = the number of minutes Gia drove. Then $2G + G + (2G - 10) = 515$. If you let G = the number of *hours* Gia drove, then an equation would be $2G + G + (2G - \frac{1}{6}) = 8\frac{7}{12}$.

c. Dien 120 miles, Gia 240 miles, and Minh 115 miles Let D = the number of miles Dien drove. Then $D + 2D + 115 = 475$.

6. a. $672,000. A drawing may help. (Let x = the number of dollars in sales for the previous year. Then $1.25x = 840,000$.)

previous year $x

last year

b. First year $75,000, and second year $105,000. (Let f = first year sales ($). Then $f + (f + .40f) = 180,000$.)

c. First year $300,000; second year $360,000; third year $540,000. Let x = first year sales in dollars. Then $x + 1.2x + 1.8x = 1,200,000$.

7. a. Susan 25 cookies, Rachel 20 cookies, and Steve 27 cookies. Let r = the number of cookies Rachel ate. Then $r + (r + 0.25r) + (1.25r + 2) = 6 \cdot 12 = 72$.

b. Rayann 3 eggs, Nell 9 eggs, and Bell 4 eggs. Let N = the number of eggs eaten by Nell. Then $\frac{1}{3}N + N + (\frac{1}{3}N + 1) = \frac{2}{3} \cdot 2 \cdot 12 = 16$.

c. 4 inches per minute. Be sure that all the values are in the same unit. Let r = Amy's rate of eating, in inches per minute. Then $(2 \cdot 4) + (1\frac{1}{3} \cdot 12) + r \cdot 3 = 36$.

d. Steve $1\frac{3}{4}$ pints (or $\frac{7}{8}$ quart), and Susan $2\frac{1}{4}$ pints (or $1\frac{1}{8}$ quarts). (Keep the units in mind. If x = the number of pints Steve drank, then $x + (x + \frac{1}{2}) = 4$. If x = the number of *quarts* Steve drank, then $x + (x + \frac{1}{4}) = 2$.)

8. $7\frac{1}{2}$ minutes. Keep in mind the units used. Let t = the time in hours they had walked. Then $6t - 4t = \frac{1}{4}$. If you let the time m be in minutes, then Carita's speed is $\frac{6}{60} = \frac{1}{10}$ mile per minute, and Jorge's is $\frac{4}{60} = \frac{1}{15}$ mile per minute. Then the equation would be $\frac{1}{10}m - \frac{1}{15}m = \frac{1}{4}$, giving the final answer directly with somewhat more difficult arithmetic.

Answers for Chapter 13: A Quantitative Approach to Algebra and Graphing

Supplementary Learning Exercises 13.1

1. a. The amount of money that Joella raises from her sister is related to the number of miles that she runs.

    b. The total amount that Joella raises from pledges is related to the number of miles that she runs. Her grand total is twice the amount that she raises from actual pledges.

    c. The number of lemons, or the amount of sugar, or the amount of water needed each depends on the number of gallons of lemonade you make.

2. a. As the cost of an airline ticket increases, the number of miles increases. This is not strictly true because of specials or lack of competition.

    b. As the number of party guests increases, the number of cans of soda bought increases.

    c. Except possibly for large amounts, the cost per pound does not depend on the number of pounds bought.

    d. As the diameter of a circle increases, so does the radius.

    e. The base and the height of a rectangle do not depend on each other. As one increases, the other can increase, stay the same, or decrease.

    f. As the width of a rectangle from the set increases, the height of the rectangle decreases.

    g. As the number of miles driven increases, the amount of gasoline used increases.

3. a. Because the light from the lightning is so fast, you sense it almost instantly. So the time for the sound is essentially the whole 5 seconds. $5 \times 343 = 1715$ meters.

    b. $d = 343 \times t$, where d is the distance in meters and t is the time between seeing the flash and hearing the thunder, in seconds.

4. a. (1) The number of people attending and the money raised through concert tickets sold; $T = 40P$ where T is the total dollars raised and P is the number of people attending.

    (2) The number of raffle tickets sold (N) and money raised by the raffle (R); $R = 10N - 300$.

    (3) No, because even if we know the number of people at the party we have no idea of the amount raised by the raffle.

    b. (1) the total amount of money taken in was $10a + 5c$ where a stands for the number of adults and c stands for the number of children who attended. There are no additional quantitative relationships that can be illustrated with an algebraic expression.

5. a. A table helps clarify this situation.

Month	Built	Sold	Total Sold
1	20	0	0
2	30	20	20
3	30	30	50
4	30	30	80
5	40	30	110
6	40	40	150

After 6 months he will have built and sold 150 units. He still has 600 units to build and sell. It will take 600 ÷ 40, or 15, additional months for all units to be built and sold, for a total of 21 months.

b. See the graph below.

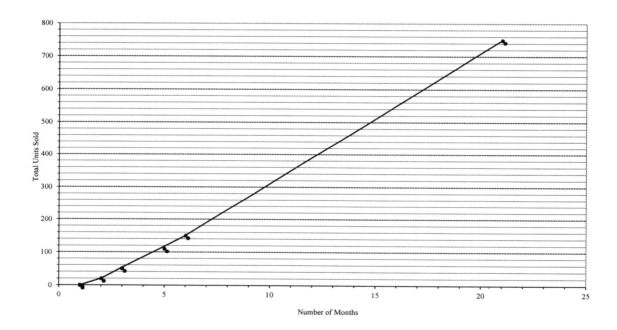

Supplementary Learning Exercises 13.2

1. a. Missing temperature: 78°; missing number of cricket chirps: 11

   b. $T = C + 39$, where T is the Fahrenheit temperature and C is the number of cricket chirps in 15 seconds.

   c. Because chirp counts will be whole numbers, strictly speaking, the points in the graph should not be joined, but they often are in order to communicate the straight-line nature of the relationship.

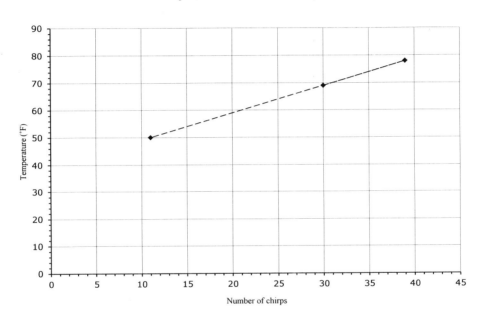

Temperature vs Number of Chirps

   d. The slope is 1.

2. a. $L = 2W$, where L is the length and W is the width of the rectangle (same units).

   b. The equation Area = (length) × (width) applies to rectangles, so taking part a into account, $A = (2W) \times W = 2W^2$.

   c. and d. If the length is on the vertical axis, your graph should go through $(0, 0)$ and have slope of 2.

3. a. Additively: The wingspan is 23 feet longer than the length (or the length is 23 shorter than the wingspan). Multiplicatively: The wingspan is $\frac{262}{239} \approx 1.1$ times as long as the length (or the length is $\frac{239}{262} \approx 0.9$ times as long as the wingspan).

   b. $81890 \div 8000 \approx 10.2$ gallons per nautical mile (sharing-equally division)

c. The graph that follows assumes a steady use of fuel, which is unrealistic because of take-offs and high-altitude cruising.

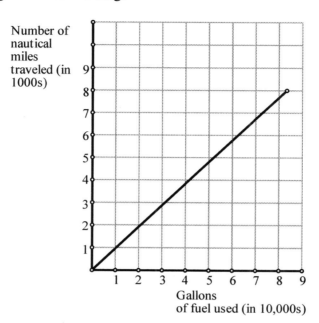

Number of nautical miles traveled (in 1000s)

Gallons of fuel used (in 10,000s)

d. $\frac{81890}{8000} \approx 10.2$

4. One possibility, with the equal slopes implying the same thickness of candle, and the steeper slope (= faster burning) implying less thickness:

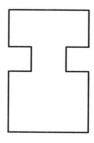

5.   a. About 1.6 kilometers
      b. About 62 miles per hour
      c. About 6.2 miles
      d. (See the graph below.)
      e. Not as steep

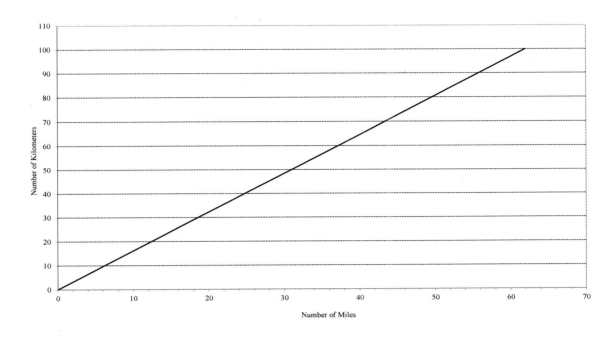

Supplementary Learning Exercises for Section 13.3

1. a. $y = 3x$, $m = 300$, $n = 120$

   b. $f(x) = 157 - 4x$, $m = 13$, $n = 50$

   c. $output = 2^{input} + 3$, $m = 1027$, $n = 11$

   d. $output = {}^-6 \cdot input$; $m = {}^-72$, $n = 100$

   e. $g(x) = x^3 + 1$, $m = 1001$, $n = 20$

   f. Notice that the lines for the x-$h(x)$ values need to be put in order. $h(x) = 4x$, $m = 100$, $n = 201$

   Each of parts (a), (d), and (f) has a rule of the form $y = mx$, with m a fixed number, so each represents a proportional relationship between x and y.

2. Only a, b, c, and d give proportional relationships between y and x. That is, only those parts give an equation that is, or can be changed to, an equation of the form $y = mx$, with m a fixed number.

3 a. Janet drove 210 miles at 50 mi/h, which took 4.2 hours, or 4 hours 12 minutes.
   Sylvia drove 240 miles at 60 mi/h, which took 4 hours. She stopped for 15 minutes, so the total time was 4 hours 15 minutes.
   So Janet arrived first, but the times are very close.

   b. The lighter line represent Janet's trip. The heavy line represents Sylvia's trip.

   c. The tent-like portion at the very beginning of her drive. The flat part on the x-axis between 30 minutes and 45 minutes.

   d. The line should begin at (0, 210) and cross over the two lines representing the trips. The line representing Janet's trip should cross it slightly before the line representing Sylvia's trip.

   e. Carlene drove 65 mi/h for 210 miles, so her trip took 3.23 hours, or about 3 hours and 14 minutes. But she left an hour later, so she will arrive about 2 minutes after Janet and about 1 minute before Sylvia. Practically speaking, the times are so close one would probably say they all arrived at the same time.

Supplementary Learning Exercises for Section 13.4

1.   a. $y = x$ b. $y = 3$ c. $y = x^2$ d. $y = 2x^2 + 2x$

x	y
0	0
2	2
5	5
8	8
9	9

Linear

x	y
0	3
1	3
2	3
6	3
10	3

Linear

x	y
0	0
1	1
2	4
3	9
4	16

Not linear

x	y
0	0
1	4
2	12
3	24
6	84

Not linear

2. Here are some possible input-output tables.

    a. $x = 3y$ b. $h = 4v^3$ c. $2y = 3x$ d. $P = 5s$

                                    is $y = \frac{3}{2}x$

y	x
0	0
2	6
4	12
6	18
8	24

v	h
0	0
1	4
2	32
3	108
10	4000

x	y
0	0
1	$\frac{3}{2}$ or $1\frac{1}{2}$
2	3
3	$4\frac{1}{2}$
4	6

s	P
0	0
1	5
2	10
3	15
6	30

Parts (a), (c), and (d) are linear because for each there is a constant rate of change. Part (b) does not have a constant rate of change and is not linear.

3. a. Let c stand for the number of cupcakes and d stand for the cost in dollars. Then $5d = c$ or $d = 0.20c$ This is a linear function. For each increase in c, d is increased by 20¢.

b. Let c stand for the amount of punch concentrate and w stand for the amount of water. Then $2c = w$, or $c = \frac{1}{2}w$. This is a linear function. For any amount of concentrate used the amount of water used is twice that amount.

4. Using an input-output table, at the beginning of the 5th year she will earn $1756.92.

Beginning of year	Salary
1	$1200
2	$1320
3	$1452
4	$1597.20
5	$1756.92

The change from year-to-year is not the same, so this would not be a linear function.

5. 1023 grains. $(2^0 + 2^1 + 2^2 + \ldots + 2^9 = 1023)$ The change from day-to-day is not the same, so this would not be a linear function.

Answers for Chapter 14: Understanding Change: Relationships Among Time, Distance, and Rate

Supplementary Learning Exercises for Section 14.1

1. The dashed segments are to help in comparing the two graphs.

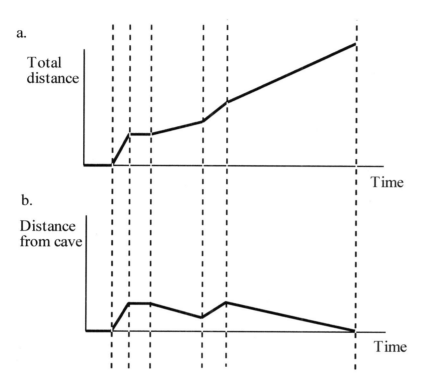

c. Alike: The first three pieces are the same because total-distance and distance-from-cave are the same. Some pieces should have the same slopes. Either graph is flat when the distance is not changing. Different: The total-distance graph never goes down, but the distance-from-cave graph can. When Wile E. heads toward the cave, the total-distance still increases but the distance-from-cave decreases.

2. a. There are many possibilities. Try to have the slopes indicate something about the speed. For example, Wile E.'s running should lead to a greater tilt than when he is walking or trotting.

b.

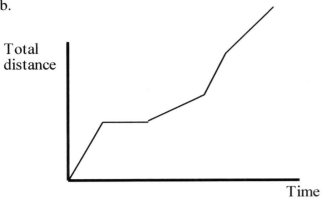

3. Your scales may be different, but your equations should be the same, of course.

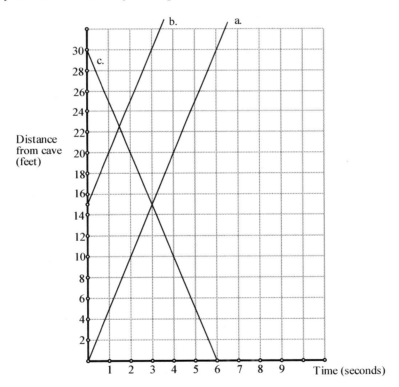

Equations: a. $d = 5t$ b. $d = 5t + 15$

c. $d = 30 - 5t$, or $d = 30 + {}^-5t$

d. Graphs a and b are parallel but meet the *d*-axis at different points. In graph c, the *d* values decrease as *t* increases. The graphs have different starting points on the *y*-axis.

e. The slopes in equations a and b are the same, and the slope in equation c appears to have the same absolute value.

4. a. The slope, 5, tells that Wile E.'s distance from the cave is increasing at 5 feet per second. The slope, 5, is the multiplier of t in the equation.

   b. The slope, 5, tells that Wile E.'s distance from the cave is increasing at 5 feet per second. The slope, 5, is the multiplier of t in the equation.

   c. The slope, ¯5, tells that Wile E.'s distance from the cave is decreasing at 5 feet per second. The slope, ¯5, is the multiplier of t in the equation.

5. There are many possibilities, but the story should start with Wile E. in the cave and then walking/strolling/running away from the cave at 10 feet per second.

6. There are many possibilities, but the story should start with Wile E. 50 feet from the cave and then moving toward the cave at 5 feet per second.

7. There are many possibilities, but your story should have Sam's speeds the same during the first and last segments, and have Sam stopped during the middle segment.

8. a. See below.   b. 5.   For part (c), $\frac{100}{18} \approx 5.56$.

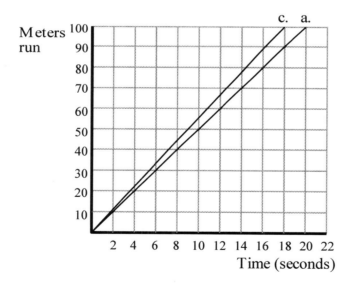

d. Any line starting at (0, 0) but going below the line for part (a)

e. It takes a little time to get to full speed and, for longer races, a runner may tire or may strategize and slow down to save strength for the end of the race.

Supplementary Learning Exercises for Section 14.2

1. a. The walker started at the motion detector and walked away from it at $\frac{1}{3}$ foot per second for 6 seconds. $d = \frac{1}{3}t$

   b. The walker started at the motion detector and walked away from it at $\frac{7}{5}$ ft/s for 5 seconds. $d = \frac{7}{5}t$

   c. The walker started 3 feet from the motion detector and walked farther away at $\frac{1}{3}$ ft/s for 6 seconds. $d = 3 + \frac{1}{3}t$

   d. The walker started 7 feet away from the motion detector and walked toward it at $2\frac{1}{3}$ ft/s for 3 seconds. $d = 7 - 2\frac{1}{3}t$ or $d = 7 + {}^-2\frac{1}{3}t$

   e. The walker started 5 feet away from the motion detector and walked toward it at 1 ft/s for 5 seconds. $d = 5 - 1t$ or $d = 5 + {}^-1t$

   f. The walker started 3 feet from the motion detector and stood there for 6 seconds. $d = 3$

2. Each of the walkers would be the same distance from his or her motion detector after a little less than 3 seconds.

3. a. The graphs with the greatest tilts (b and d) would belong to the fastest walker away from the detector (b) and toward the detector (d).

   b. The equation with the greatest positive slope ($d = \frac{7}{5}t$) would belong to the fastest walker away from the detector, and the equation with the negative slope with the greatest absolute value ($d = 7 + {}^-2\frac{1}{3}t$) would belong to the fastest walker toward the detector.

4. (Equations only are given.)

   a. $d = 4 + 2t$ or an algebraic variation, like $d = 2t + 4$ or even $d - 2t = 4$, because they describe the same relationship between the variables, algebraically.

   b. $d = 4 - 1t$ or $d = 4 - t$ or $d = 4 + {}^-1t$

   c. $d = 6 - 2t$ or an algebraic equivalent

Supplementary Learning Exercises for Section 14.3

1.

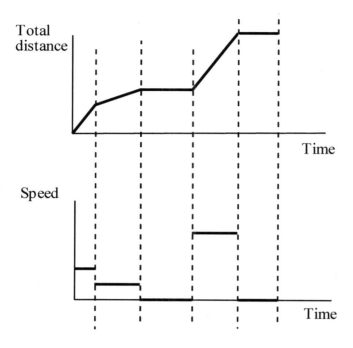

2. a.

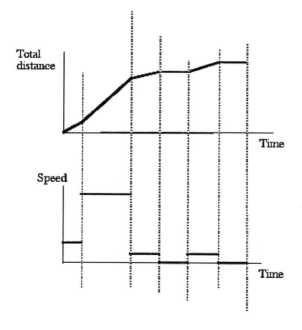

b.

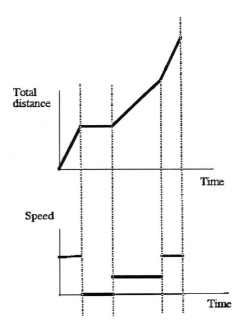

c.

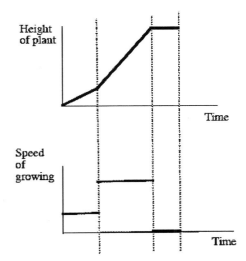

3. a. The distance stays the same while Wile E. is stopped, so the corresponding piece of the graph is horizontal.

   b. The speed should be zero, so the corresponding piece of the graph is on the time axis.

   c. The position stays the same while Wile E. is stopped, so the corresponding piece of the graph is horizontal.

4. Dots on the ends of segments may help with alignment, as for part (b) below, without intending to imply the open/filled-in dot convention mentioned earlier.

a.

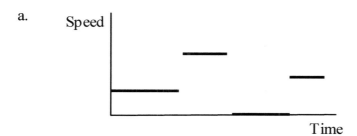

b.

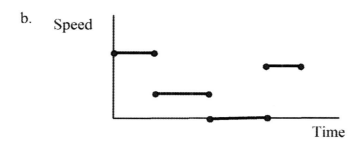

5. Here are some (rough) samples.

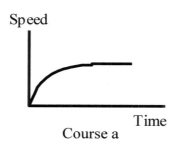

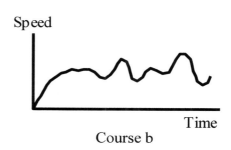

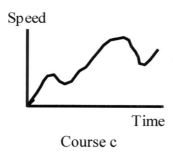

Supplementary Learning Exercises for Section 14.4

1. There are many possible stories, but yours should reflect a high rate of speed for the first piece, then a stop, then a resumption of the high speed, then a slowing down, and finally a further slowing down.

2. The speed increases steadily and then stays constant.

3. There are many possibilities, but your story should indicate that Diane sped up steadily for 20 minutes until she was biking 0.1 mile per minute, then she biked at that speed for 20 minutes, and then she again speeded up until she was biking 0.2 mile per minute.

4. a. There should be a flat piece after P while she ate/drank something at the snack shop, but there is none.

   b. S. From the vertical scale, the distance walked during S is greater than those for P and Q (and R).

   c. P. From the horizontal scale, the time taken for P is greater than those for Q, R, and S.

   d. Q. The differences in distances for Q is relatively great for the short time span that Q took.

   e. Not for certain, although the first part of the walk took a goodly amount of time for not a great distance. In particular, Q very likely did not involve any steep hills because it was her fastest piece [part (d)].

5. Type a goes with graph ii; it will fill up fast at the start, then slow down, and then speed up as it goes past the middle of the vase. Type b goes with graph i because it fills steadily all the way to the top. Type c goes with graph iv because it will fill up faster at the start and then less fast later. Type d goes with graph iii because it will fill up slowly at first but gradually speed up.

6. Your choice of proportions should reflect that because the vase fills faster and faster during the first 5 seconds, it is getting narrower and narrower. Then the vase gets slightly wider and wider.

Answers for Chapter 15: Further Topics in Algebra and Change

Supplementary Learning Exercises for Section 15.1

1. $y = 2x + 1$, from $\frac{y-5}{x-2} = \frac{9-5}{4-2}$

2. $y = 3x + 4$, from $\frac{y-7}{x-1} = \frac{22-7}{6-1}$

3. $y = {}^-x + 10$, from $\frac{y-9}{x-1} = \frac{14-9}{-4-1}$

4. $y = \frac{1}{3}x + \frac{1}{2}$, from $\frac{y-2\frac{1}{2}}{x-6} = \frac{2-2\frac{1}{2}}{4\frac{1}{2}-6}$

5. $y = 8x - 5$, from $\frac{y-3}{x-1} = 8$

6. $y = \frac{3}{4}x + 2$, from $\frac{y-5}{x-4} = \frac{3}{4}$

7. $y = {}^-4x - 7$, from $\frac{y-{}^-11}{x-1} = {}^-4$

8. $y = {}^-2x - 9$, from $\frac{y-{}^-7}{x-{}^-1} = {}^-2$

9. $y = 6x - 3$, because it will have the same slope: $6 = \frac{y-3}{x-1}$

10. $y = {}^-2x + 9$, because it will have the same slope: ${}^-2 = \frac{y-7}{x-1}$

11. The slopes using every pair of the points should be equal, but they are not: 2, 3, and $2\frac{1}{3}$.

12. The equation of the line through the two points is $y = 5x + 2$, so the line is parallel to the line $y - 5x = 7$, or $y = 5x + 7$.

13. The equation for his descent is $h = {}^-700t + 5300$, from $\frac{h-3200}{t-3} = \frac{400-3200}{7-3}$.

    a. 700 ft/min downward

    b. 5300 ft

    c. $\frac{53}{7} \approx 7.6$ min after jumping (from $0 = {}^-700t + 5300$)

14. 230 ft, from $t = 0$ in $\frac{d-500}{t-6} = 45$

Supplementary Learning Exercises for Section 15.2

1. a.   Each equation will be true if you substitute $x = 5$ and $y = 4$ into them.

   b.   Yes. The substitution leads to $22 = 22$ and $4 = 5 - 1$, both true.

   c.   Their graphs will have the point (2, 7) in common.

2. a.   The equations in the system do not have a common solution. Graphically, it means their graphs never meet or cross.

b. One of the equations can be obtained from the other by multiplying both sides of the other by the same nonzero number. Graphically, it means that their graphs are the same sets of points.

3. $x = 5, y = 8$, or more compactly, $(5, 8)$

4. $(10, 3)$

5. $(1, {}^-2)$

6. The system is inconsistent.

7. $s = 4, t = 2\frac{1}{2}$

8. $(3, 1)$

9. $({}^-2, {}^-3)$

10. $(25, 21)$

11. The equations are dependent (multiply the second equation by 2 and reorganize the equation).

12. $({}^-5, {}^-6)$

13. a. Any linear equation with slope 5 but not passing through $(0, 0)$

 b. Any equation obtained by multiplying both sides of $y - 5x = 19$ by the same nonzero number

14. a. Any equation that gives a slope of $\frac{20}{42} = \frac{10}{21}$ but does not have a y-intercept at ${}^-\frac{163}{42}$.

 b. Check that $(2, 3)$ is a solution of each of your equations. At the start, work gradually—for example, starting with $2x - y$, you could finish by seeing what $2x - y$ equals when $x = 2$ and $y = 3$. $2x - y = 2 \cdot 2 - 3$, so $2x - y = 1$ would be one equation.

15. Jones $75, Smith $250, and Quan $225

16. Misha and Lavonne each scored 16 points, and Sue scored 8.

17. Misha, 16 points; Sue, 10 points; and Lavonne, 5 points

18. Angela, $13.20; Marcy; $4.40; and Lucia, $39.60

19. The two elevators were at the same height after $1\frac{1}{4}$ min. The problem did not ask what the height was then, but you could calculate it by substituting $t = 1\frac{1}{4}$ into either of your equations.

20. The first time after $\frac{2}{3}$ min, the second time (when A is stopped) $2\frac{1}{3}$ min after the start of the trips, and the third time $5\frac{2}{3}$ min after the start of the trips.

Supplementary Learning Exercises for Section 15.3

1. a.   A-One starts out 30¢ behind but gains 3¢ each $\frac{1}{18}$ mile, so in $10\frac{1}{18}$-mile pieces (= $\frac{5}{9}$ mile), A-One's fare will be the same as Speedy's: $2.70.

   b.   The scale for the distance is best expressed as $\frac{1}{18}$-mile pieces. Because the fare jumps at each $\frac{1}{18}$-mile segment, joining points to get the cost isn't actually proper but is often done. If the graphs crossed at some distance that is not an exact multiple of $\frac{1}{18}$, one could read an incorrect answer from the graph.

   c.   You may have used other letters for the variables, of course. Speedy: $c = 1.50 + 0.12d$, with c being the cost of the fare in dollars and d being the number of $\frac{1}{18}$-mile distances, or $c = 150 + 12d$, with c being the cost of the fare in cents. For A-One: $c = 1.20 + 0.15d$ (or $c = 120 + 15d$). Equating the two costs gives $1.50 + 0.12d = 1.20 + 0.15d$, or $0.30 = 0.03d$, or $10 = d$. $10 \times \frac{1}{18} = \frac{10}{18} = \frac{5}{9}$ mile, with each company's fare being $2.70.

2. a.   Perfecto starts out $10 behind but gains $2 every 15 minutes, so until five 15-minute periods of work have been done, Perfecto would be less expensive than Pronto. Longer times (more than 75 minutes or 1.25 hours) would favor Pronto.

   b.   The scale for time is best expressed in terms of 15-minute pieces. As in Exercise 1, because the charge jumps at the 15-minute marks, the points should not actually be joined (but often are) because reading the graph for, say, a 20-minute job could give an incorrect cost.

   c.   Perfecto: $c = 30 + 20p$, where c = the cost for a job that takes p 15-minute periods. Pronto: $c = 40 + 18p$. From $30 + 20p = 40 + 18p$, one gets $p = 5$, so the costs are equal then. Trying a value less than 5 for p shows that Perfecto is less expensive than Pronto. (Notice that this finding is clear from the graph.)

3. a.   The male starts out 210 calories ahead, in calories used, but for every hour of walking, the female uses 30 calories more. So in 7 hours of walking at 3 miles per hour (21 miles!), the two would use the same number of calories, 3970.

   b.   Deciding on the scales may be a key part. Expressing the calories used scale in terms of 100s and the "distance" in terms of the number of hours (= 3 miles) might be best. Joining the points here is correct because there is no requirement to walk for a whole number of hours.

c.   Female: $C = 1520 + 350h$, where C = the number of calories and h the number of hours walked. Male: $C = 1730 + 320h$. $1520 + 350h = 1730 + 320h$ gives $h = 7$, so 7 hours of walking, or 21 miles, will result in the two using the same number of calories.

4.   It can be interesting to share your reactions with those of others.

5.   The graphs should intersect at ($^-40$, $^-40$), so when either temperature is $^-40$, so is the other one.

Supplementary Learning Exercises for Section 15.4

1.   No. She has 2.8×108 = about 302.4 grade points, and with a 4.0 for the last 12 semester hours, she will earn 48 more grade points, for a total of about 350. But for a 3.0 average for all 120 hours, she needs 360 grade points. The best she can get is about $350.4 \div 120 = 2.92$.

2.   a. The total time for the trip is $7\frac{1}{3}$ hours, so the average speed for the whole 400 miles is
$400 \div 7\frac{1}{3} = 54\frac{6}{11} \approx 54.5$ mph.

b. This time the total time for the trip is $9\frac{5}{7}$ hours, giving an average for the whole trip of
$400 \div 9\frac{5}{7} = 41\frac{3}{17} \approx 41.2$ mph.

c. The 65 mph for the whole 400 miles would take $6\frac{2}{13}$ hours. The 50 mph trip going takes 4 hours, so that leaves only $2\frac{2}{13}$ hours for the return 200 miles, requiring a speed of
$200 \div 2\frac{2}{13} = 92\frac{6}{7}$, or about 93 mph!

3.   Needed is information about either the number of classes at each school, enabling one to find the total enrollment for both schools and the total number of classes. Alternatively, if there are assurances that there is the same number of classes at each school, then the given 27 as an average is correct.

4.   Without knowing whether the same number of tickets was sold in each city, it is risky to use the $54 figure. If, say, there were 200 tickets sold in the first city (yielding $9600) and 1000 tickets in the second city (yielding $60,000), the average for the two cities is $(9600 + 60000) \div (200 + 1000) = 69600 \div 1200 = 58$ dollars, not $54.

5.   a. There is no information about how many weeks the two people dieted.

b. The 2.5 figure would be accurate if the two people dieted for the same number of weeks.

6.   a. Turtle must run at 8 m/s to tie. (Rabbit takes $[30 \div 20] + [30 \div 5] = 7.5$ seconds in all, so Turtle must run the 60 meters in 7.5 seconds to tie. $60 \div 7.5 = 8$.

b. Turtle must run at $60 \div 11 = 5\frac{5}{11}$ m/s to tie.

c. Turtle must run at 16 m/s to tie.

d. Turtle must run at $60 \div 30\frac{1}{2} = 1\frac{59}{61} \approx 2$ m/s.

7. a. In each part, Rabbit's average speed is the same as Turtle's speed because that steady speed gives the same distance as the Rabbit trips over and back.

   b. None would be changed. (Think of the 120 as two 60 m races.)

8. Rabbit takes $(40 \div 10) + (40 \div 16) = 6.5$ seconds to go over and back. Turtle must cover only 60 m, having a 20-m head start, so Turtle must go $60 \div 6.5 = 9\frac{3}{13} \approx 9.2$ m/s, to tie.

Supplementary Learning Exercises for Section 15.5

1. a. Given a value for sales, one can predict the profit (in large part).

   b. Given the speed, the stopping distance is predictable.

   c. Given the amount of rainfall, the agricultural yield is predictable.

   d. Given a measure of the care for the product, its life can be predicted.

2. a. $p(n) = 6n + 2$. One justification: The first shape has perimeter 8 units. Each of the $n-1$ additional ones adds only 6 units because the two that overlap disappear. All together that gives $8 + 6(n-1)$, or $6n + 2$, units.

   b. $c(n) = 2.98n + 0.06(2.98n)$, or $1.06(2.98n)$, or $3.1588n$. Justification: Each box costs $2.98, so n boxes cost $2.98n$, and then the sales tax is added on.

   c. $p(n) = 8n + 2$. One justification: The top four of the first shape are repeated n times for $4n$ units, as are the bottom four for $4n$ more, plus the two at the ends. Total: $4n + 4n + 2$. Try a justification similar to that for part (a).

   d. $p(n) = 10n + 2$. One justification: The top five in Shape 1 are repeated n times, as are the bottom five, plus the two at the ends. Total: $5n + 5n + 2$. Again, try a justification similar to that in part (a).

   e. $c(n) = 3.95n + 12n(0.05)$, or $4.55n$. Justification: The packs cost $3.95 each, and because each of the 12 cans in a pack requires a 5¢ deposit, that adds $12(0.05)$ to the cost for each pack.

3. a. 17   b. 51

   c. 7   d. 21

4. a. 2025   b. 144   c. $56\frac{1}{4}$ d. 547.56

5. a. No. For an input of x, Machine 1 gives $2x + 8$, but Machine 2 gives $2(x + 8)$.

   b. 142.8. Machine 1 gives 2(27.7) + 8 = 63.4, and then Machine 2 doubles 63.4 + 8 = 71.4, giving 142.8.

   c. 150.8. Machine 2 doubles 27.7 + 8 = 35.7, giving 71.4, and then Machine 1 doubles 71.4, giving 142.8, and adds 8: 150.8.

6. $f(x) = x^2 + 4$

7. $g(x) = x^3 - 1$

8. $y = 15x - 4$

9. $y = 2x^2$

Part III: Reasoning About Shapes and Measurement

Answers for Chapter 16: Polygons

Supplementary Learning Exercises for Section 16.1

1.   a. Trapezoid   b. Heptagonal region

      c. Equilateral triangle or regular triangle (These are the best answers, but isosceles triangle also applies.)

      d. Square (This is the best answer, but rectangle, rhombus, kite, parallelogram also apply.)

      e. Rhombus (This is the best answer, but parallelogram and kite also apply—*diamond* is not commonly used in geometry.)

      f. Right-triangle region

      g. Isosceles trapezoid

      h. Hexagon region

2.   a. Your shape should have 5 sides, with at least two of different lengths.
      b. Do you have 12 sides?

      c. Not possible; a square is both equilateral and equiangular and therefore regular.

      d. Any rectangle, even a square, will do.

      e. Not possible; you probably know that a rectangle has four 90° angles.

      f. Not possible, for then the angle sum would be greater than 180°.

      g. Any equilateral triangle will do.

      h. The right angles should be at the "sides" of a kite in the common kite drawing.

3.   a. 90°   b. 80°   c. $360 - (90 + 80 + 76) = 114°$ d. 66°   e. 104°
      f. 110°   g. 64°   h. 120°   j. $360 - 60 - 110 - 64 = 126°$

      k. 145°   m. 104°

      n. The interior angle by n has $720 - (90 + 145 + 104 + 118 + 119) = 144°$, so 36°.

4.   a. $(9 - 2) \cdot 180 = 1260°$ b. $(12 - 2) \cdot 180 = 1800°$ c. $(m - 2) \cdot 180°$

5. Samples:

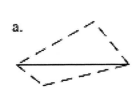

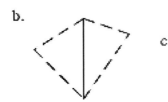

6. Samples:

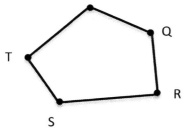

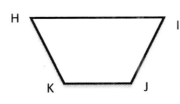

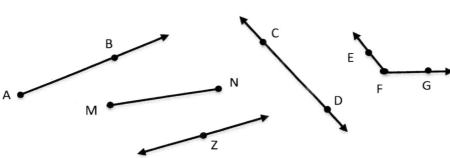

Supplementary Learning Exercises for Section 16.2

1. a. Yes   b. Yes   c. No   d. Yes   e. No

    f. No   g. Yes   h. No   i. No

2. a. Yes   b. No   c. Yes   d. No   e. Yes (squares)

Supplementary Learning Exercises for Section 16.3

1. a.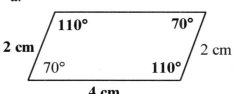

   b.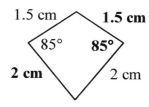

   c.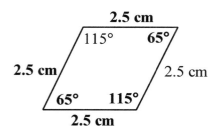

2. a. A trapezoid that is not at all an isosceles trapezoid should give a counterexample.

   b. Use a rectangle that is much, much longer than it is wide.

   c. Your trapezoid for part (a) should give a counterexample.

   d. Any rectangle gives a counterexample.

3. a. Your measurements should support the conjecture.

   b. Again, your measurements should support the conjecture.

   c. The evidence is based on only one example of a rectangle. Results might potentially be different for some other rectangle(s).

Answers for Chapter 17: Polyhedra

Supplementary Learning Exercises for Section 17.1

1. a. 8 faces, 12 vertices, 18 edges

    b. 5 faces, 6 vertices, 9 edges

    c. 6 faces, 8 vertices, 12 edges

    d. 6 lateral edges; 3 lateral edges

2. b. 6 faces (Count each flat surface of the whole shape, not the individual square regions.), 12 edges

3.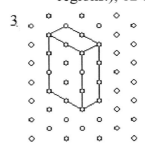

4.   Front view   Right view   Top view

Supplementary Learning Exercises for Section 17.2

1. a. C, D, E, F   b. A   c. A, C   d. A, G   e. F
2. a. 100   b. 50   c. 51   d. 51
3. a. 20   b. 20   c. 40   d. 22
4. No
5. a. 45   b. 45   c. 46   d. 46
6. No
7. a. 12   b. 60   c. 120
8. a. 14   b. 26   c. 14
9. a. 150   b. 60   c. 48

Supplementary Learning Exercises for Section 17.3

1. You should have a pentagonal prism.

2. a. You should have a hexagonal pyramid, with the top vertex off-center with the base.

b. You should have a quadrilateral pyramid, with the top vertex off-center with the base.

c. Your drawing for part (a) should suffice.

3. a. Your net should have 6 triangles and a hexagon. The side of a triangle that matches the "next" triangle should be the same length. At least one triangle should not be isosceles.

b. Your net should have 4 triangles and a quadrilateral. The side of a triangle that matches the "next" triangle should be the same length. At least one triangle should not be isosceles.

c. Your net for part (a) should work.

4.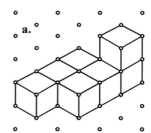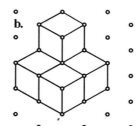

5. Only shape (d) can be obtained from the starting shape. Notice in the starting shape where the square corner is.

6. a. A regular hexagonal region and isosceles triangular regions.

b. Rectangular regions

c. A rectangular region and triangular regions, probably isosceles

d. Rectangular regions and isosceles trapezoidal regions

7. The diagram will look like the one for quadrilaterals, but with "prisms" attached to each type of quadrilateral.

8. Think hierarchically (every rhombus is a special trapezoid, so the result should be the same). Or, each of the faces in each case is a quadrilateral, so the sum of all the angles in all the faces will be the same for any quadrilateral prism, 2160°.

Supplementary Learning Exercises for Section 17.4

1. It is not possible for any prism to be congruent to a given pyramid because it would be impossible to make the figures match exactly. A prism has at least 3 faces that are parallelogram regions, and a pyramid can have at most one such face.

2. Congruent shapes can be matched exactly, so each face in one prism must correspond to a face in the other prism.

3. Yes. Some edge in prism Y must match the 4-in. edge in prism X exactly.

4. The matching edges will have the same lengths (2 cm, 6 cm, or 3 cm), as will the matching angles. Matching faces will be the same size.

5. One possibility is the following prism. Be sure that your version is oblique and that the hidden edges are shown.

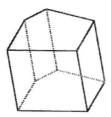

Supplementary Learning Exercises for Section 17.5

1. a. The student counted all the visible edges and vertices, not realizing that the unfolding of the polyhedron would give duplicate edges and multiple copies of some vertices in many places in the net.

    b. Take into account that the unfolding of the polyhedron will use some edges twice and some vertices more than once. Try to anticipate which ones are counted more than once.

2. The base of the Great Pyramid is a square region, not an equilateral triangle region like the other faces.

3. Answers will vary.

4. Try to be systematic. For example, with 4 squares in a row, how can you add the other 2 squares to get a net for a cube? With 3 squares in a row? If you get stuck, try Cube Nets at http://illuminations.nctm.org.

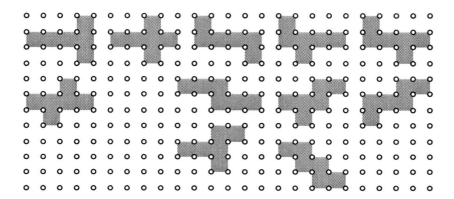

Answers for Chapter 18: Symmetry

Supplementary Learning Exercises for Section 18.1

1.   a. Only one reflection symmetry, with the line of symmetry
     going through the midpoints of the top and bottom of the
     polygon.

   b. 2 reflection symmetries (in the dotted lines) and 2 rotational
      symmetries (180° and 360°, center at highlighted point).

   c. Only one reflection symmetry, in a line along the horizontal segment midway between
      the two other horizontal segments.

   d. 3 reflection symmetries (see drawing for the reflection
      lines) and 3 rotational symmetries (120°, 240°, and
      360°, center at highlighted point.

   e. One reflection symmetry in a line through the
      highlighted point and the midpoint of the segment
      between the 2 "antennas." No nontrivial rotational
      symmetries.

2.   Check your answers with someone else. You may get some ideas for Exercise 1.
3.   Your argument can use a 60° rotational symmetry or use reflections in 2 lines of
     symmetry that are not diagonals.
4.   Answers vary by individual.
5.   The arrangement of the two halves is also important.

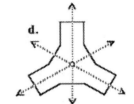

Supplementary Learning Exercise for Section 18.2

1 and 2.
   a. There are two reflection symmetries and two rotational symmetries (180° and 360°).
      See the drawing below for parts of the cross-sections and the (heavy) axis of rotation
      sticking out toward you.
   b. Same as for part (a) except the axis of rotation is vertical. See the drawing below.

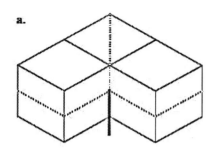
a.

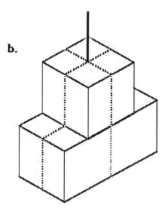
b.

3. There are 7 reflection symmetries, with 6 of the planes of symmetry suggested by the lines of symmetry in the hexagonal bases and with the 7th plane of symmetry cutting through the lateral faces, halfway down from the top base and perpendicular to the lateral edges.

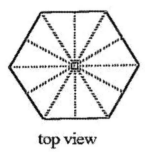
top view

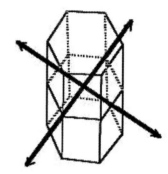

There is a grand total of 12 rotational symmetries. There are 6 rotational symmetries using an axis joining the centers of the two hexagonal bases (60°, 120°, 180°, 240°, 300°, and 360°). There are an additional 6 rotational symmetries from axes that are lines of symmetry for the hexagonal cross-section from the 7th plane of symmetry; each has 180° (the trivial 360° symmetry has already been counted).

4. Did you actually find four planes of symmetry?

5. No axis of symmetry exists, nor does any plane of symmetry. The slanted nature of shape B and the non-rhombus nature of the parallelogram faces keep pieces from matching when rotated in possible axes or reflected in possible planes of symmetry.

Answers for Chapter 19: Tessellations

Supplementary Learning Exercises for Section 19.1

1. For the triangles and the trapezoid, notice that if you can get a parallelogram region you can clearly tessellate the plane with the polygon.

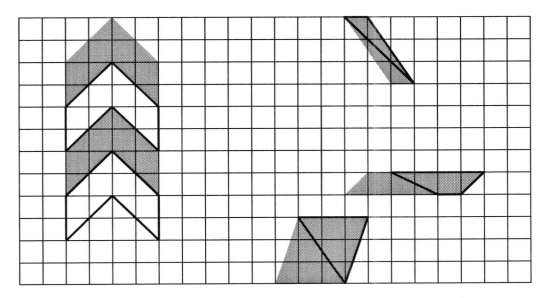

2. Probably. For example, for the chevron shape a reflection in a line down the middle of the part of the tessellation that is shown gives a symmetry; although there do not appear to be any rotational symmetries for the chevron in the one given (yours might have some 180° rotational symmetries).

3. Another regular pentagon would have to fit into either question-marked space. But there is not room for two more (216°) in the first case and not enough (108° + 180° = 288°, leaving 360° − 288° = 72°) in the second case.

4. $x + x + x = 360$, so $x = 120$ degrees.

5. Yes, each of the five tetrominoes can tessellate the plane.

6. Yes. See, for example, the tessellation in Learning Exercise 9(a), or this shape:

Supplementary Learning Exercises for Section 19.2

1. Can   2. Cannot (the cylinder-shaped kind)   3. Cannot   4. Can

5. Cannot   6. Can   7. Can   8. Cannot

Answers for Chapter 20: Similarity

Supplementary Learning Exercises for Section 20.1

1. The sizes of each pair of corresponding angles are the same. The ratio of the lengths of corresponding sides all have the same value (the scale factor).

2. In each part, the image is a line segment 2.5 times as long as the original segment.

3. Only the location of the images would be different.

4. a. The angles in the image will all be 90°, so the image is a rectangle. The scale factor is $\frac{7}{4} = 1\frac{3}{4}$ (remember to use measurements from the center), so the new dimensions are $\frac{7}{4} \times 28 = 49$ cm and $\frac{7}{4} \times 56 = 98$ cm.

    b. The image will still be a rectangle, but the scale factor is now $\frac{4}{7}$, giving new dimensions for the rectangle of $\frac{4}{7} \times 28 = 16$ cm and $\frac{4}{7} \times 56 = 32$ cm.

5. $a, b = 118°$ $c = 45°$ $d = 17°$ (scale factor = 1.5) $y = 22.5$ cm $x = 12$ cm

6. $c = a = 98°$ $b = 60°$ $d = 22°$ (s.f. = 3.5) $x = 6$ m $y = 14$ m

7. $a = 83° = c$ $b = e = 41°$ $d = 139°$ (s.f. = 1.2) $x = 24 - 20 = 4$ ft $y = 25$ ft

8. $a, c, f = 70°$ $b, d, e, g = 110°$ $v = 24$ cm (s.f. = $\frac{2}{3}$) $z, y = 36$ cm

    $x = 12$ cm $w = 8$ cm

9. a. $18   b. $30   c. $18   d. $30
    e. 40 lb   f. 32 lb   g. 40 lb   h. 32 lb

Supplementary Learning Exercises for Section 20.2

1. The pairs of triangles in each of parts (a), (c), and (d) are similar.

2. $b = 125°$ $a, c, d = 90°$ $e = 55°$ (s.f. = 0.6) $y = 7.2$ m $x = 4.8$ m $z = 9$ m

3. $a = 111°$ $b = c = 93°$ $d = 75°$ $e = 81°$ (s.f. = 1.2) $y = 10.8$ in. $z = 9.6$ in. $x = 4.1$ in.

4. Ratio of areas (= s.f.^{2}) and ratio of perimeters (= s.f.), so #2, 0.36 and 0.6; #3, 1.44 and 1.2.

5. a. Regular pentagon

    b. The perimeter of the second polygon, which has 5 sides of the same length, is given to be 45 in., so each side is 9 in. Then the scale factor from larger to smaller is $\frac{9}{12} = \frac{3}{4}$.

    c. So, using part (b), the scale factor from larger to smaller is $\frac{9}{12} = \frac{3}{4}$.

Supplementary Learning Exercises for Section 20.3

1. The points of the three pyramids can be matched so that corresponding angles have the same size and so that the ratio of the lengths of every pair of corresponding sides is the same.

2. No. The lengths may not have the same ratio, and the angles need not be the same size.

3. a. No. The ratios $\frac{4}{6} = \frac{2}{3}$ and $\frac{3}{5}$ are not equal.

   b. Yes. The angles are all right angles, so their sizes are equal, and each of $\frac{12}{9}$, $\frac{20}{15}$, and $\frac{8}{6}$ equals $\frac{4}{3}$.

   c. No. Although $\frac{10}{20}$ and $\frac{12}{24}$ each equals $\frac{1}{2}$, $\frac{14}{21} \neq \frac{1}{2}$.

   d. Yes. The angles are all right angles, so their sizes are equal, and putting the lengths in order for each prism, $\frac{6}{9} = \frac{8}{12} = \frac{10}{15}$ because each equals $\frac{2}{3}$.

4. a. 2200 in. ≈ 184 ft, the actual length of the shuttle, although $2200 \div 12$ gives 183 ft. 4 in.

   b. 1 : 1,000,000

5. a. 22 cm by 33 cm by 44 cm

   b. 37 cm by 55.5 cm by 74 cm

   c. 12.5 cm by 18.75 cm by 25 cm

6. a. Areas (larger compared to smaller), $(2\frac{1}{5})^2 = 4\frac{21}{25}$, so the larger shape has $4\frac{21}{25}$ times the area of the smaller shape; volumes, $(2\frac{1}{5})^3 = 10\frac{81}{125}$

   b. Areas, 13.69, volumes, 50.653

   c. Areas, 156.25%; volumes, 195.3125%

7. a. The triangles' sides: 57.6 cm, 50.4 cm, and 64.8 cm; height 21.6 cm

   b. Sides of triangles: 6.8 cm, 5.95 cm, and 7.65 cm; height 2.55 cm

   c. Sides of triangles: 12.8 cm, 11.2 cm, and 14.4 cm; height 4.8 cm

8. a. Areas, $7.2^2 = 51.84$, so the larger shape has 51.84 times the area of the smaller shape; volumes, $7.2^3 = 373.248$

   b. Areas, 0.7225; volumes, 0.614125

   c. Areas, 256%; volumes, 409.6%

9. a. $\frac{320}{5} = 64 = 4^3$, so the scale factor is 4.

   b. Using the scale factor from part (a), the ratio of the areas is $4^2 = 16$. Polyhedron Y has 16 times as much area as polyhedron X has.

Answers for Chapter 21: Curves, Constructions, and Curved Surfaces

Supplementary Learning Exercises for Section 21.1

1. a. A segment of the circle

    b. A sector of the circle

    c. A central angle

    d. An inscribed angle

    e. A (minor) arc

2. Your diameter, a line segment, should go through the center of the circle and have its endpoints on the circle. Your radius, a line segment, should have endpoints at the center of the circle and at a point on the circle. For parts (c) and (d), compare Exercise 1(a) and (b). Your chord, a line segment, should have endpoints on the circle and (in this case) not go through the center of the circle. Your central angle can be any angle with its vertex at the center of the circle. Your inscribed angle should have its vertex on the circle, with chords as its sides. Your minor arc should be any (connected) piece of the circle that is less than half the circle.

3. a. 70° (It is a central angle intercepting an arc of 70°.)

    b–d. 35° (They are inscribed angles intercepting an arc of 70°.)

    e–g. 90° (They are inscribed angles intercepting half the circle because of the diameter and thus intercepting an arc of 180°.)

4. a. and b. You can check the accuracy of your work with a protractor.

    c. and d. The angle should be 90° or close to it. So a reasonable conjecture is that if you bisect two angles with outside sides making a straight line, the two bisectors make an angle of 90°.

    e. *Hint:* A straight angle has 180°.

5. Your quadrilateral should look like a rhombus. Use the rotational symmetry of the rectangle and a line of symmetry to justify your conjecture.

6. You may have bisected any angle of the pentagon and argued that the bisector will be a line of symmetry (how can you be sure that the other parts do coincide when reflected? The last step in the argument is difficult.). Or you might have constructed the perpendicular bisector of a side of the pentagon and argued that it is a line of symmetry. (How can you be sure that the other parts do coincide when reflected? The last step in the

argument is difficult.) You could also construct the perpendicular to a side from the opposite vertex, but it is more difficult to argue that this perpendicular is a line of symmetry.

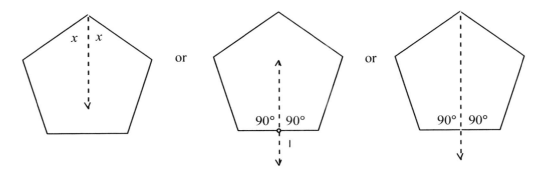

In any case, your argument should be about the properties your constructed line and the regular pentagon have rather than an argument based on direct measurements or folding the shape on the line.

Supplementary Learning Exercises for Section 21.2

1. a. Your great circles should appear to meet each other at the ends of a diameter.
    b. Your cone should be a right circular cone.

    c. The vertex of your cone should not be right over the center of the base.

    d. Start with a curve such as the one in Figure 1 in Section 21.2.

2. For example, look at your hand (as well as things outside your body).

3. Something like the drawing in Learning Exercise 18(a) in Section 25.2, although you may have put in the hidden edge of the base and some other curves or line segments.

4. a. To fill the rest of the container requires another $\frac{1}{4}$ of the container. Because $2\frac{1}{2}$ cups fill $\frac{3}{4}$ of the container, $\frac{1}{3}$ of $2\frac{1}{2}$ cups would fill the other $\frac{1}{4}$ of the container. $\frac{1}{3} \times 2\frac{1}{2} = \frac{5}{6}$ cup to fill the rest of the container. (A drawing might help.)
    b. $2\frac{1}{2} + \frac{5}{6} = 3\frac{1}{3}$ cups for the whole container (or from another way of thinking, $4 \times \frac{5}{6} = 3\frac{1}{3}$ cups), so $\frac{1}{2} \times 3\frac{1}{3} = 1\frac{2}{3}$ cups for half the container.

Answers for Chapter 22: Transformation Geometry

Supplementary Learning Exercises for Section 22.1

1. a. Translation, rotation, reflection (Could you name them without referring elsewhere?)

    b. ...is a movement that does not change lengths or angle sizes.

2. The lengths of the sides of the image will be the same as those in the original (2 cm, 2.5 cm, 4.45 cm, and 3 cm), and the angle sizes will be the same also (90°, 148°, 54°, and the 68° size not given in the original).

3. Decisions might be based on eye-sight, key ideas, or orientation as an aid.

    original→ A, reflection   original→ B, reflection

    original → C, rotation   original→ D, translation

    original → E, rotation

4. The same orientation for original and each of images (C), (D), and (E). The opposite orientation for original and each of images (A), (B), and (F).

5. Parts (a)–(d) below. (e) Each image is congruent to the original, because they are related by a rigid motion.

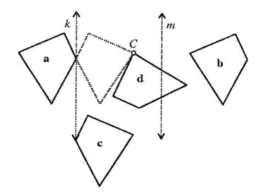

6. B. 180 degree rotation of A; C. Translation; D. Reflection of A; E. Glide-reflection; F. Translation; G. Reflection; H. Translation

Supplementary Learning Exercises for Section 22.2

1. Usually, you can tell from your work whether the image looks correct and is in the correct place.

2. Only the first two points of the images are labeled.

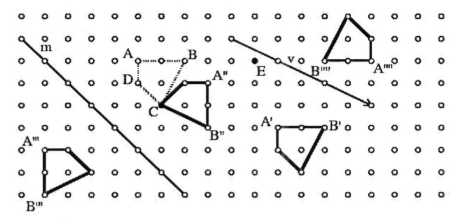

3. Here is a sample exercise with answers. None of the vertices in the images are labeled. Can you label them with confidence?

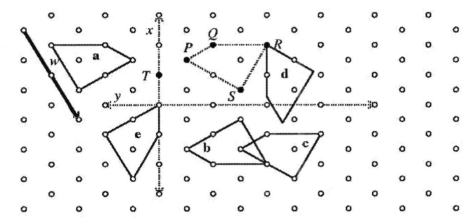

4. Yes. Each has the effect of leaving the point where it was initially, no matter what the intermediate movement was.

5. Each describes the same movement: move 3 cm straight to the right.

Supplementary Learning Exercises for Section 22.3

1. If you have doubts about your results, have someone else look at them.

2.

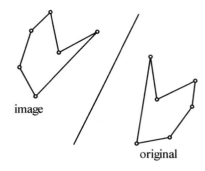

3.

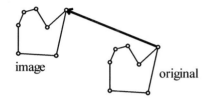

4.

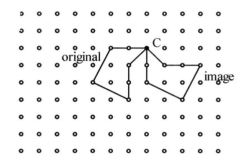

Center C,
90Þ counter-clockwise

5.   Rotation with center at the common vertex, angle 120° counterclockwise (or 240° clockwise).

6.   a. and b.

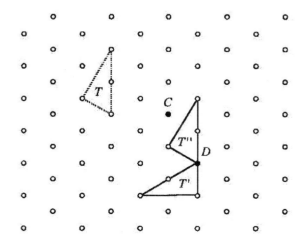

    c. A translation, 3 diagonal spaces in a roughly south-easterly direction

Supplementary Learning Exercises for Section 22.4

1. Translation, rotation, reflection, and glide-reflection (Did you answer without referring elsewhere?)

2. The translation

3. a. Translation 2 in. north

   b. Translation 2 in. north Notice that order does not matter with translations.

   c. Rotation 10° counterclockwise, center C

   d. Rotation 10° counterclockwise, center C (Notice that order does not matter with rotations that have the same center.)

4. a. Given → (A) by a translation (Can you draw the vector?)
   b. Given → (B) by a rotation

   c. Given → (C) by a rotation

   d. Given → (D) by a glide-reflection

   e. Given → (E) by a reflection (Can you locate the line of reflection?)

5. a. Did you do the translation first?

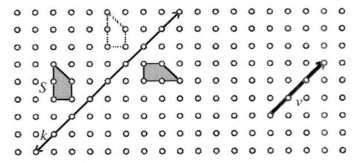

   b. A glide-reflection

   c. Even though the intermediate step (from the reflection) would give a different location for the image, the final image would be exactly the same.

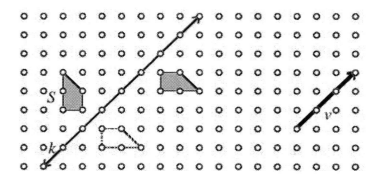

6.   Each leaves every point in its same location. (They describe the same *function*.)

Supplementary Learning Exercises for Section 22.5

1.   a. Translation symmetries, by moving the shape to the right or left by (multiples of) the distance from the left-hand start of a semicircle to the right-hand end of a triangle. Reflection symmetries with lines of reflection perpendicular to the line the figures are resting on and through the top of a semicircle or a bottom vertex of a triangle. (There are no rotational or glide-reflection symmetries.)

   b. Translation symmetries, by moving the shape to the right or left by (multiples of) the length of the side of the triangle. Reflection symmetries with lines of reflection perpendicular to the line the triangles are resting on and through a vertex of a triangle. Reflection symmetries with lines of reflection perpendicular to the line the triangles are resting on and through the point of intersection at the base of the triangles. (There are no rotational or glide-reflection symmetries.)

   c. Translation and reflection symmetries as in part (b). Rotational symmetries of 180° with centers at either vertex of the triangle on the line the triangles are resting on. Glide-reflection symmetries from reflecting in the line the triangles are resting on and translating by (multiples of) the side of the triangle to the right or left.

2.   Answers will vary.

3.   The final image will be a rectangle, 15 cm by 20 cm. We can't say anything about its location or orientation without knowing more about its original orientation and the location of point *C*.

4.   Did you see rotational and reflection symmetries? Did you see a possible tessellation with the shape? Each of these involves transformations.

Answers for Chapter 23: Measurement Basics

Supplementary Learning Exercises for Section 23.1

1. More than 7 units. It takes more of a smaller unit to equal a certain number of a larger unit.

2. You do not know whether the "glasses" for you and your friend were the same size.

3. a. and b. Measurements, except for counts, are only approximations. Eyesight and especially limits of the measuring tool mean measurements should not be considered "exact."

4. a. Shortest: $2\frac{1}{2}$ units; longest: just under $3\frac{1}{2}$ units
   b. Shortest: $3\frac{1}{4}$ units; longest: just under $3\frac{3}{4}$ units

5. a. Greater than 100. A meter is shorter than a kilometer, so it will take more meters to make a given number of kilometers.
   b. Less than 100. A kilometer is longer than a meter, so it will take fewer kilometers to make a given number of meters.
   c. Greater than 100. An ounce is smaller than a pound, so it will take more ounces to make a given number of pounds.
   d. Less than 100. A pound is greater than an ounce, so it will take fewer pounds to make a given number of ounces.

6. a. 4   b. 30   c. $\frac{5}{8}$ (0.6 is sometimes used)   d. 8
7. a. 0.49   b. 120   c. 49 000   d. 0.049
   e. 520   f. 5.2

8. a. 512 megabytes = 512 000 000 bytes
   b. 4 gigabytes = 4 000 000 000 bytes
   c. 2 terabytes = 2 000 000 000 000 bytes

9. a. 3 lb 13 oz (because 1 pound = 16 ounces)
   b. 3.7 kg (Notice that the basic relationships in SI mesh well with decimal arithmetic.)
   c. 2 yd 2 ft 7 in (because 1 yd = 3 ft and 1 ft = 12 in.)
   d. 3.58 m (Again, notice how easy the arithmetic is with metric units.)
   e. 1 gal 3 qt (because 1 gallon = 4 quarts)
   f. 2 hours and 45 minutes

Supplementary Learning Exercises for Section 23.2

1. Answers will vary.
2. a. About 5 cm   b. About 10 cm or 1 dm   c. About 3 cm
3. The measurements appear to have been made to the nearest half inch. So the $3\frac{1}{2}$ in. could be as small as $3\frac{1}{4}$ in. or as long as up to $3\frac{3}{4}$ in., and the $4\frac{1}{2}$ in. could be as small as $4\frac{1}{4}$ in. or as long as up to $4\frac{3}{4}$ in. So the sum of the lengths might actually be as small as $3\frac{1}{4} + 4\frac{1}{4} = 7\frac{1}{2}$ in. or as long as up to $3\frac{3}{4} + 4\frac{3}{4} = 8\frac{1}{2}$ in. Notice that the potential "errors" add up.

4. a. 0.056   b. 560   c. 8.4   d. 0.084   e. 4500   f. $1\frac{1}{4}$
5. a. 5' 6" or 66"   b. 5.5 m   c. 14.28 cm   d. 25.48 cm   e. $34\frac{2}{3}$ yd
6. Good estimates would be in the vicinity of 135°, 75°, 170°, and 30°. "Good" might mean within 10° or 5°, depending on your experience.

7. Answers will vary.

8. a. 69°   b. 39° 43'   c. 66° 46' 48"   d. 169°   e. 75°, 105°, 105°
   f. 104°, 76°, 76°   g. 70°, 96.5°   h. 135° each   i. 156° each   j. 40°, 40°

9. *a, b, e, f* = 139°   *c, d, g* = 41°   *h, k, l, o* = 69°   *j, i, m, n* = 111°
10. $b = 104°$ $a = 360 - (104 + 136) = 120°$ $c = 68°$ $d = 60°$
    $e = 180°$ $f = 40°$ $g = 180°$ $h = 90°$ $i = 50°$ $j = 100°$
    $k = 64°$ $l = 168 - 52 = 116°$ $m = 116°$ $n = 96°$ $p = 116°$
    $q = 2 \cdot 50 - 56 = 44°$ $r = 2 \cdot 88 - 56 = 120°$ $s = 140°$ $t = 92°$ $u = 130°$

11. A whole (such as the circle) can be measured by cutting it into parts, measuring each of those, and adding those measures. For example, $360 = 136 + 104 + a$, so you could calculate *a*.

12. a. 42   b. 28

Answers for Chapter 24: Area, Surface Area, and Volume

Supplementary Learning Exercises for Section 24.1

1. An $8\frac{1}{2}$ in. by 11 in. sheet has area about 6 dm^2 (using metric sense and envisioning dm^2 on the sheet), so 200 pages would have an area of about 1200 dm^2.

2. a. One J is $\frac{3}{4}$ as large as a K, so 30 J = $22\frac{1}{2}$ K. One J is $\frac{3}{5}$ as large as an L, so 30 J = 18 L .
   b. Similarly, 1 K = $1\frac{1}{3}$ J and 1 K = $\frac{4}{5}$ L. Hence, 30 K = 40 J = 24 L.
   c. As before, 1 L = $1\frac{2}{3}$ J and 1 L = $1\frac{1}{4}$ K, giving 30 L = 50 J = $37\frac{1}{2}$ K.
   d. From the above relationships, 17 J = $12\frac{3}{4}$ K = $10\frac{1}{5}$ L.
   e. As before, 17 K = $22\frac{2}{3}$ J = $13\frac{3}{5}$ L.
   f. 17 L = $28\frac{1}{3}$ J = $21\frac{1}{4}$ K
   g. m J = $\frac{3}{4}m$ K = $\frac{3}{5}m$ L
   h. n K = $\frac{4}{3}n$ J = $\frac{4}{5}n$ L
   i. p L = $\frac{5}{3}p$ J = $\frac{5}{4}p$ K

3. A drawing suggests that 1760 rows with 1760 yd^2 in each row make a square mile, so $1760 \times 1760 = 3,097,600$ yd^2.

4. 1 mm^2 = 0.01 cm^2 (10 rows, with 10 mm^2 in each row make 1 cm^2)

5. a. 0.78   b. 1500
   c. 140   d. 14 000 (or 14,000)
   e. 1.28   f. 260

6. "Surrounding" the polygon or a part of its region with a rectangle may help.

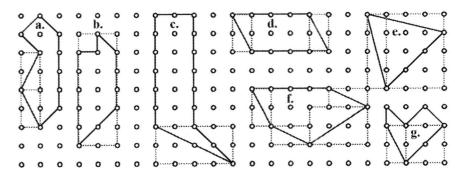

   a. 8 units   b. $8\frac{1}{2}$ units   c. 14 units   d. 8 units

e. $16 - (2 + 2 + 4\frac{1}{2}) = 7\frac{1}{2}$ units   f. 11 units   g. $4\frac{1}{2}$ units

7.   a. One way: top-bottom views, 2 units each; right-left faces, 6 units each, front-back, 4 units each. Total: 24 units

b. One way: top-bottom views, 13 units each; right-left views, 9 units each; front-back views, 11 units each. Total: 66 units

8.   a. $1.2^2 \cdot 72 = 103.68$ cm^2

b. $\frac{72}{\text{smaller area}} = 1.2^2$, and so smaller area $= 72 \div (1.2)^2 = 50$ cm^2

9.   The ratio of the areas is 1:9, so the scale factor could be 1:3 (or, if the original and final were switched, 3:1).

10.   a. $\left(\frac{36}{4}\right)^2 = 81$ (or 81:1)

b. There are many values that have an 81 to 1 ratio (for example, 162 and 2, 243 and 3, 810 and 10, 90 and $1\frac{1}{9}$, 180 and $2\frac{2}{9}$, and so forth).

11. a. Double each answer in Supplementary Learning Exercise 6: a-16, b-17, c-28, d-16, e-15, f-22, g-9

b. Divide each answer in Supplementary Learning Exercise 6 by 8: a-1, b-$1\frac{1}{16}$, c-$1\frac{3}{4}$, d-1, e-$\frac{15}{16}$, f-$1\frac{3}{8}$, g-$\frac{9}{16}$

Supplementary Learning Exercises for Section 24.2

1.   a. 2000   b. 2000   c. 49

d. 1300   e. 4 000 000 (or 4,000,000)   f. 1000

2.   a. $4 \times 12^3 = 6912$ b. 1   c. 81

3.   a. $x > y$. The dm^3 unit is smaller than the m^3 unit, so it takes more of them to equal a certain number of the larger unit.

b. $x < y$. The dm^3 unit is larger than the cm^3 unit, so it takes fewer of them to equal a certain number of the smaller unit.

c. $x < y$. The ft^3 unit is larger than the in^3 unit, so it takes fewer of them to equal a certain number of the smaller unit.

d. $x > y$. The ft^3 unit is smaller than the yd^3 unit, so it takes more of them to equal a certain number of the larger unit.

4.   a. 7 cubes   b. 31 cubes

c. 7 mL and 31 mL, because 1 cm^3 = 1 mL

d. 7 ft^3 and 31 ft^3. Because 1 yd^3 = 27 ft^3, the part (b) polyhedron has volume greater than 1 yd^3.

5. With luck, you would have some cubes of the same size at hand (and perhaps a small empty box). You might say, "The volume of something is the number of these cubes that it would take to fill it, or to make a shape just like it. Let's see what the volume of this box is."

6. a. First shape: S.A. = 20 units, V = 5 cubes

    Second shape: S.A. = 68 units, V = 24 cubes

    b. Image of first: S.A. = $6^2 \times 20 = 720$ units, V = $6^3 \times 5 = 1080$ cubes

    Image of second: S.A. = $6^2 \times 68 = 2448$ units, V = $6^3 \times 24 = 5184$ cubes

7. a. S.A. = $2^2 \times 856 = 3424$ cm^2, V = $2^3 \times 1.68 = 13.44$ L
    b. S.A. = $(\frac{2}{3})^2 \times 856 = 380\frac{4}{9}$ cm^2, V = $(\frac{2}{3})^3 \times 1.68 = \frac{8}{27} \times 1.68 = \frac{8}{9} \times 0.56 = \frac{4.48}{9} \approx 0.498$ L

    c. S.A. = $10^2 \times 856 = 85\ 600$ cm^2 (or 85,600 cm^2), V = $10^3 \times 1.68 = 1680$ L

    d. S.A. = $0.1^2 \times 856 = 8.56$ cm^2, V = $0.1^3 \times 1.68 = 0.00168$ L (or 1.68 mL)

Answers for Chapter 25: Counting Units Fast: Measurement Formulas

Supplementary Learning Exercises for Section 25.1

1. The statement means that pi (π) cannot be expressed exactly as a terminating or repeating decimal, or equivalently, in the form $\frac{\text{integer}}{\text{nonzero integer}}$.

2. The date March 14 is often 3/14, resembling 3.14, an approximation for π.

3. For a rectangle, $A(\text{trapezoid}) = \frac{1}{2}h(a+b)$ becomes $A(\text{rectangle}) = \frac{1}{2}w(l+l) = lw$.

    For a rhombus, $A(\text{trapezoid}) = \frac{1}{2}h(a+b)$ becomes $A(\text{rhombus}) = \frac{1}{2}h(b+b) = bh$.

4. With 3.14 as an approximation for π, the diameter of the ball is given by $28.5 \approx 3.14d$, or $d \approx 28.5 \div 3.14 \approx 9.08$ in., so two balls at one time would take more than the 18 in. available.

5. a. 38.936 m   b. 132 in.   c. $12\pi + 24$, or about 61.7 cm
    d. 15.4π, or about 48.38 cm (*semi*-circles)   e. $33 + 16.5\pi$, about 84.84 cm

6. a. $\pi 6.2^2$, about 120.8 m^2 b. 1386 in^2 c. $144 + 36\pi$, or about 257.1 cm^2
    d. $\frac{1}{2}\pi 4^2 + \frac{1}{2}\pi 5^2 + \frac{1}{2}\pi 6.4^2 + \frac{1}{2}8 \cdot 10 = $ about 80.98 cm^2

7. 49π is about two times 25π. Your cut out circular region should make the result reasonable.

8. a. $220 + 73\pi$, or about 449.3 cm^2 b. $240 - 144 = 96$ cm^2 c. $360 - 180 = 180$ cm^2

9. a. 7.2 m   b. 16 in.   c. 3 cm

10. a. 2537.16 cm^2 b. 208 m^2

11. 950.4 cm^2 (The lateral faces are rectangular regions.)

12. a. 266π, or about 835.7 in^2 b. 1120π, or about 3518.6 cm^2
    c. (radius = 4 cm) 88π, or about 276.5 cm^2 d. (radius = 9 cm) 342π, or about 1074.4 cm^2

13. $2(220 + 73\pi) + (11\pi + 5\pi + 32)5$, or $600 + 226\pi$, or 1310 cm^2

14. From the description, the two boxes are similar with scale factor 1.5. So, *S.A.* of larger box $= 1.5^2 \cdot 236 = 531$ in^2. $V = 1.5^3 \cdot 240 = 810$ in^3.

15. Either $A = \pi a^2$ or $A = \pi b^2$. As the two values change to the same number, the ellipse becomes a circle, so the "new" formula is no surprise.

16. a. $\frac{12 \times 18}{9} = 24$ sq. yd; 24   \$15.60 = \$374.40

b. $\dfrac{216}{13.1} \times 1.05 \approx 17.3$ bundles, so must buy 18 bundles at $13.1 \times \$2.99 = \$39.17$ per

bundle or $705.06

c. Allowing for door, window, and closet openings, the area of the walls is $416\frac{8}{9}$ sq. ft, so a gallon might not be enough. If you buy 2 gallons, $66.

Supplementary Learning Exercises for Section 25.2

1. In your sketch, the two bases should be congruent, and their area, B, should be indicated. The height, h, should be the distance between the bases, not the length on a slant.

2. An area formula should give units in, say, cm^2, but $A = 4\pi r^3$ would give cm^3. Similarly, a volume formula should give units in cm^3, but $V = \frac{4}{3}\pi r^2$ would give cm^2.

3. a. 448 ft^3 b. 400 m^3 c. 3600 cm^3 d. 600 ft^3 e. 216 in^3
   f. $33\frac{1}{3}$ cm^3 g. 1620π, or 5089.4 cm^3 h. 67.5π, or 212.1 in^3

   i. The volume of each cone would be one-third that of the cylinder.

4. The new volume is the old volume multiplied by the cube of the scale factor. Here, then, (a) new $V = 4^3 \cdot 448 = 28672$ ft^3; (c) new $V = 4^3 \cdot 3600 = 230400$ cm^3; and (h) new $V = 4^3 \cdot 67.5\pi = 4320\pi \approx 13572$ in^3.

5. If you are using the usual cubic units, then each length should be expressed in the same unit. Here, the 1.1 m should be replaced by 110 cm, if the answer is to be in cm^3. Notice that paying attention to units might have helped this student avoid the error.

6. a. $V = 2304\pi$, about 7238.2 in^3; S.A. $= 576\pi$, about 1809.6 in^2
   b. V about 8.2 cm^3; S.A. about 19.6 cm^2

   c. V about 2094.4 m^3; S.A. $= 300\pi$, or about 942.5 m^2

   d. $V = 2250\pi$, about 7068.6 ft^3; S.A. $= 675\pi$, about 2120.6 ft^2

   e. $V = \frac{2}{3}\pi r^3$; S.A. $= 3\pi r^2$

7. Volume, $571666\frac{2}{3}\pi$, or 1 795 943.8 ft^3; S.A. (not including the bases of the cylinder) 19600π, or 61 575.2 ft^2

8. a. 23 cm   b. 88 cm   c. 24 $in.^2$ d. ($r = 2$) 12π, or 37.7 cm^2

9. The small b is for the length of a side of the parallelogram, whereas the capital B is for the area of the base of the prism or cylinder.

10. a. 325,851.429

b. About 815

c. Metric calculations (using liters and ares or even hectares) would involve dealing only with powers of 10.

Answers for Chapter 26: Special Topics in Measurement

Supplementary Learning Exercises for Section 26.1

1. a. 90 cm
    b. $18 + \sqrt{164}$, about 30.8 cm
    c. 112 m
    d. $28 + \sqrt{224}$, about 43 cm

2. a. 56" or 4' 8"
    b. 40 cm
    c. $14 + \sqrt{98}$, about 23.9 cm
    d. About 40 m
    e. $8 + 8\pi$, about 33.13 cm

3. a. 192 in^2
    b. 96 cm^2
    c. 24.5 cm^2
    d. 100 m^2

4. $\sqrt{16 + 100} = \sqrt{116} \approx 10.8$ You might ask the student to check his/her answer to see whether $14^2 = 116$ (it doesn't, of course).

5. The triangles in parts (a) and (c) are right triangles. For part (a),
    $40^2 + 42^2 = 1600 + 1764 = 3364 = 58^2$. For part (c),
    $28^2 + 96^2 = 784 + 9216 = 10000 = 100^2$, and 100 cm = 1 m.

6. Design A: volume = 1600 m^3;

    outside S.A (not including floor) = $550 + 5\sqrt{244} + 6\sqrt{200} \approx 713 \, \text{m}^2$

    Design B: volume = about 467 m^3

    outside SA (not including floor) = $270 + 2.5\sqrt{674} + 3.5\sqrt{650} \approx 424 \, \text{m}^2$

7. S.A.: $h^2 + 2^2 = 2.9^2$ gives $h = 2.1$ m. So the area of each of the two bases of the right triangular prism is $\frac{1}{2} \cdot 4 \cdot 2.1 = 4.2 \, \text{m}^2$. The area of the floor is 24 m^2, and the area of each rectangular slanted side is 17.4 m^2. The total is $4.2 + 4.2 + 24 + 17.4 + 17.4 = 67.2 \, \text{m}^2$.

    Volume: With the area from above, $V = 4.2 \times 6 = 25.2 \, \text{m}^3$.

8. a. 7
    b. 5
    c. 2a
    d. 3d
    e. 5
    f. $\sqrt{41} \approx 6.4$
    g. 13
    h. 5

    i. $\sqrt{(m-p)^2 + (n-q)^2}$

9. $4 + 5 + \sqrt{58} + 7 = 16 + \sqrt{58} \approx 23.6$

Supplementary Learning Exercises for Section 26.2

1. Most people recognize that building larger and smaller units by powers of 10 and communicating the different sizes by the same prefixes are two strong features. That their definitions are scientifically based is another strong feature. Disadvantages might be lack of familiarity for many in the United States and the existence of many everyday objects that are not based on the metric system but need repair in a metric world.

2. One way of thinking is $\frac{100 \text{ years}}{\text{century}} \cdot \frac{365 \text{ days}}{\text{year}} \cdot \frac{24 \text{ hours}}{\text{day}} \cdot \frac{50 \text{ miles}}{\text{hour}} = 43{,}800{,}000 \, \frac{\text{miles}}{\text{century}}$.

3. a. 400 mg (1 carat is 200 mg)   b. 75% (24 karat gold is pure gold)

4. Ignoring a difference in taxes and possible differences in cost from the suppliers, we know that the Canadian (Imperial) gallon is larger than the U.S. gallon.

5. The area of the room is 180 ft^2, or 20 yd^2. So the carpet cost would be 20×12.99, or about $260.

6. One way of thinking is to note that $\frac{\text{number of births}}{\text{population in 1000s}}$ = the birth rate. Hence, $\frac{\text{population in 1000s}}{\text{number of births}} = \frac{1}{\text{birth rate}}$, or population in 1000s = $\frac{\text{number of births}}{\text{birth rate}}$. This equation gives:

    a. Population in 1000s in 2005 = $\frac{4138349}{14.0} \approx 295,596$, so the 2005 population was about 295,596,000 (or just 260 million).

    b. About 250,240,000   c. About 202,790,000

Part IV: Reasoning About Chance and Data

Answers for Chapter 27: Quantifying Uncertainty

Supplementary Learning Exercises for Section 27.1

1. a. Not probabilistic. This has happened and is either true or false. One rephrasing is, "The probability that a person born next year will grow up in the same town that he or she was born in."

    b. Probabilistic, if read as referring to a future situation.

    c. Not probabilistic, because it has already happened. One rephrasing is, "The probability you will go to the grocery store on some Friday in the future, given circumstances like the ones you are now in."

    d. Probabilistic, if it refers to a future situation with the same kind of deck of cards.

2. a. 0.5   b. 0   c. 1.0   d. 0.03   e. 0.85

3. a. There are 12 outcomes: H1, H2, H3, H4, H5, H6, T1, T2, T3, T4, T5, and T6.

    b. H1, H3, H5

    c. Outcomes are the simplest things that can result with an experiment (as in part (a)), whereas an event is *any* collection of outcomes (as in part (b)), even a collection with only one outcome or even none. Hence, a single outcome can be considered an event.

4. The probability derives from the large number of tickets purchased. This is the fraction of tickets that would win.

5. Possible outcomes for 2 cards drawn from a regular deck of 52 cards consist of any two cards: jack of diamonds & queen of spades; 3 of clubs and 10 of hearts.

    Examples of possible events for 2 cards drawn from a regular deck of 52 cards: drawing two black cards; drawing two face cards; drawing a pair.

Supplementary Learning Exercises for Section 27.2

1. a. $120° + 120° + 90° = 330°$, so B must have an angle $360° - 330°$, or $30°$. Hence, $P(A) = P(D) = \frac{120}{360} = \frac{1}{3}$, $P(C) = \frac{90}{360} = \frac{1}{4}$, and $P(B) = \frac{30}{360} = \frac{1}{12}$.
    Alternatively, $P(B) = 1 -$ sum of the other probabilities, or
    $P(B) = 1 - (\frac{1}{3} + \frac{1}{3} + \frac{1}{4}) = 1 - \frac{11}{12} = \frac{1}{12}$.

    b. $P(B) = \frac{1}{12}$ means that if you spin the spinner a large number of times, you would get B on about $\frac{1}{12}$ of the spins.

2. No, a probability cannot be negative. As the fraction $\frac{\text{number of outcomes in event}}{\text{total number of trials}}$, a probability cannot be less than 0 or greater than 1. Events cannot happen a negative number of times.

3. a. $P(J) = \frac{90}{360} = \frac{1}{4}$. $360° - 90° = 270°$, so the angle size for each of K and L is $270° \div 2 = 135°$. Hence, $P(K) = P(L) = \frac{135}{360} = \frac{3}{8}$. Alternatively, $P(K)$ and $P(L)$ must share equally the $1 - \frac{1}{4} = \frac{3}{4}$ left from $P(J)$, and $\frac{3}{4} \div 2 = \frac{3}{8}$.

    b. $P(L) = \frac{3}{8}$ means that if you spin the spinner a large number of times, you would get L on about $\frac{3}{8}$ of the spins.

4. a. Letting p be the probability of one of the first three outcomes, $p + p + p + \frac{1}{2}p = 1$, so $\frac{7}{2}p = 1$, and $p = \frac{2}{7}$. The first three have probability of $\frac{2}{7}$ and the last is $\frac{1}{7}$. To find the sizes of the angles, multiple 360° by each probability (getting $102\frac{6}{7}°$ for each of the largest three angles and $51\frac{3}{7}°$ for the fourth).

    b. Let $P(A) = x$. Then $P(B) = \frac{1}{2}x$, and $P(C) = \frac{1}{4}x$. The sum of the three probabilities must be 1, so $x + \frac{1}{2}x + \frac{1}{4}x = 1$. Then $\frac{7}{4}x = 1$, or $x = \frac{4}{7}$. That is, $P(A) = \frac{4}{7}$, $P(B) = \frac{2}{7}$, and $P(C) = \frac{1}{7}$. To find the sizes of the angles, multiply 360° by each probability (getting $205\frac{5}{7}°$, $102\frac{6}{7}°$, and $51\frac{3}{7}°$).

    c. Letting x and y be the two probability values, we have $3x + 2y = 1$ from the sum of all five probabilities and $x + y = \frac{3}{8}$ from the information about the two values. Solving those two equations (double each term in the second equation and subtract the result from the first equation), we get $x = \frac{1}{4}$ and $y = \frac{1}{8}$. Your spinner should have three 90° sectors (for A, B, and C) and two 45° ones (for D and E).

    d. Letting x and y be the two probability values, we again have $3x + 2y = 1$ from the sum of all five probabilities but $3x = 2y$ from the information about the two values. Substituting for $2y$ in the first equation gives $3x + 3x = 1$, or $x = \frac{1}{6}$. From $3 \cdot \frac{1}{6} = 2y$, we get $y = \frac{1}{4}$. (Alternatively, substitute $2y$ for $3x$ in the first equation, giving $y = \frac{1}{4}$ and then $x = \frac{1}{6}$.) The sectors for A, B, and C should have 60° and the two for D and E should have 90°.

5. a. Since each part is the same size the probability for each is part is $\frac{1}{4}$. The probability of winning a prize is $\frac{3}{4}$.

    b. $P(\text{large prize}) = \frac{1}{4}$

    c. $P(\text{no prize}) = \frac{1}{4}$

    d. The area of each section is $\frac{1}{4} \cdot 120 = 30 \text{ in}^2$.

    e. The probability can be found by dividing the area of the section by the total area.

6. a. $\dfrac{\sqrt{3}}{60}$ b. $1 - \dfrac{\sqrt{3}}{60}$

7. It may be helpful to complete the division lines.

a. $P(X) = \frac{1}{9}$ b. $P(Y) = \frac{3}{9} = \frac{1}{3}$ c. $P(Z) = \frac{5}{9}$

8. Drawings can vary greatly. For part (a) the regions for A and B should each be the same size, and the region for C twice that size. For part (b), the region for E should be half the size of each of the other four regions. Here are two ways for part (b); notice that the probabilities are relatively easy to see from the drawings:

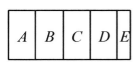

9. a. If the list of students is known for the class or if the class is visible, this probability would be theoretical. One would calculate the number of females compared to the total number in the class.

    b. Theoretically. Find all of the ways that two dice can come up (36) and count how many of those sum to seven (6 ways—1, 6; 6, 1; 2, 5; 5, 2;.3, 4; and 4, 3). The (theoretical) probability is $\frac{6}{36} = \frac{1}{6}$.

10. a. Expect to spin R 750 times, from $\frac{150}{360} \cdot 1800$. The 150 comes from $360 - (90 + 120)$.

    b. Expect to spin S 450 times.

    c. Expect to spin T 600 times.

    d. Yes, it is possible.

11. It depends. Suppose there are x red marbles and y blue marbles. The probability of drawing a red is then $\frac{x}{x+y}$. Putting in one more of each color changes the probability to $\frac{x+1}{x+y+2}$. How do these two fractions compare? Rewrite with a common denominator:

    $\frac{x(x+y+2)}{(x+y)(x+y+2)}$ and $\frac{(x+1)(x+y)}{(x+y)(x+y+2)}$, or $\frac{x^2+xy+2x}{\text{denominator}}$ and $\frac{x^2+xy+x+y}{\text{same denominator}}$. Comparing the numerators leads to comparing $2x$ and $x + y$, or just x and y. If $x = y$, the probabilities are the same. If $x > y$, then the first fraction (from before the marbles are added) is greater than the second fraction, so the probability has decreased when the marbles were added. If $x < y$, then the first fraction (from before the marbles are added) is smaller, so the probability has increased when the marbles were added. If you considered examples, be sure that you included all the cases: $x = y$, $x > y$, and $x < y$.

12. a. This event has already occurred, so either it was a boy or not. Can you rephrase the situation to make it probabilistic?

    b. This question does not refer to probability. You either have a dog or do not. This could also be rephrased to ask a probability question.

13. a. The odds of drawing a red queen are 1 to 25.

    b. The odds of that chicken laying an egg with a double yolk are 125 to 875 or 1 to 7.

    c. The odds of Jim hitting the bull's eye are 2 to 3.

14. a. The probability of Jon making a base hit is $\frac{3}{7}$.

    b. The probability of a calico cat being male is $\frac{1}{100,000}$ or 0.00001.

    c. The probability of being put on hold when calling the doctor is $\frac{23}{25}$.

15. a. 14/36 or 7/18   b. 8/30 or 4/15   c. $P(YY) = 1/36$ or zero without replacement

Supplementary Learning Exercises for Section 27.3

1. a. The area of the purple is $\frac{3}{10}$. The simulation could use a random number list with this code:

                   0, 1 = red
                   2, 3, 4 = green
                   5, 6 = yellow
                   7, 8, 9 = purple

    b–e. Your answers to these parts will depend on your data. The experimental probabilities may be close to the theoretical, but should vary slightly. The answer to part (e) should be 1 – your answer to part (d).

2. Your simulation might be based on sets of 4 two-digit numbers, with, say, 01–65 being effective, and 66–00 being ineffective. Your probability from the simulation should be close to 0.179. The more repetitions, the closer the estimate will be to this value.

3. One way to simulate this is to use five playing cards with consecutive numbers—for example, use a 2 (for B), 3(for I), 4 (for N), 5 (for G), and 6 (for O). Then shuffle and blindly pull the cards out one at a time. A draw of 23456 would be a B-I-N-G-O. Another would be to use a random number list and code the letters carefully. Many repetitions will be needed to get a nontrivial answer.

4. a. Any 35% simulation will do—for example, random numbers.

    b–d. Results will vary. With lots of repetitions, your result for part (c) should be around 0.005, and your result for part (d) should be around 0.03.

5. The simulation should indicate snoring 59% of the time and include a group size of four. There are assumptions being made in the question itself. For example, there is no distinction between males and females snoring, but males do snore more often than

females. Also, we are not considering whether snoring runs in families. With lots of repetitions of your simulation, your result should be around 0.12.

Answers for Chapter 28: Determining More Complicated Probabilities

Supplementary Learning Exercises for Section 28.1

1.

2. a. Your spinner should have $0.6 \times 360 = 216$ degrees for heads.

b. Your tree diagram should give these outcomes and probabilities:

HH, 0.36   HT, 0.24   TH, 0.24   TT, 0.16

c. Here are the corresponding probabilities for an honest coin:

HH, 0.25   HT, 0.25   TH, 0.25   TT, 0.25

3. a. RRR   WRR   BRR
    RRB   WRB   BRB
    RWR   WWR   BWR
    RWB   WWB   BWB   Each outcome has probability $\frac{1}{12}$.

b. Only outcome RRR has all the same color, so the probability is $\frac{1}{12}$.

c. 10 outcomes have one or more reds, so the probability is $\frac{10}{12} = \frac{5}{6}$.

d. $1 - \frac{5}{6} = \frac{1}{6}$, using part (c), or by directly counting the outcomes with no reds (WWB and BWB).

e. RWB, WRB, WWR, BRB, and BWR have exactly one red, so the probability is $\frac{5}{12}$.

4. a. $\left(\frac{1}{2}\right)^{12} = \frac{1}{4096}$ (Think of going through the tree diagram.)
    b. If you repeated the 12 tosses many, many times, about $\frac{1}{4096}$ of the repetitions would have all tails.

5. a.

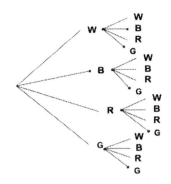

b. WW   BW   RW   GW

WB   BB   RB   GB

WR   BR   RR   GR

WG   BG   RG   GG

c. | Outcome | Prob | Outcome | Prob | Outcome | Prob | Outcome | Prob |
|---|---|---|---|---|---|---|---|
| WW | $\frac{1}{36}$ | BW | $\frac{1}{36}$ | RW | $\frac{1}{18}$ | GW | $\frac{1}{18}$ |
| WB | $\frac{1}{36}$ | BB | $\frac{1}{36}$ | RB | $\frac{1}{18}$ | GB | $\frac{1}{18}$ |
| WR | $\frac{1}{18}$ | BR | $\frac{1}{18}$ | RR | $\frac{1}{9}$ | GR | $\frac{1}{9}$ |
| WG | $\frac{1}{18}$ | BG | $\frac{1}{18}$ | RG | $\frac{1}{9}$ | GG | $\frac{1}{9}$ |

6. a.

1, 1	2, 1	3, 1	4, 1	5, 1
1, 2	2, 2	3, 2	4, 2	5, 2
1, 3	2, 3	3, 3	4, 3	5, 3
1, 4	2, 4	3, 4	4, 4	5, 4
1, 5	2, 5	3, 5	4, 5	5, 5

b. Because each section has the same angle, each outcome has probability $\frac{1}{25}$.

7. a. Spinning a spinner twice is similar to drawing with replacement.

b. Sample space of products of two spins

1	2	3	4	5
2	4	6	8	10
3	6	9	12	15
4	8	12	16	20
5	10	15	20	25

c. $\frac{2}{25}$

d. $\frac{16}{25}$

8. a. Draw 3 balls twice with replacement

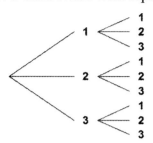

   b. Draw 3 balls twice without replacement

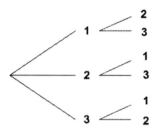

   c. Draw 3 balls three times with replacement

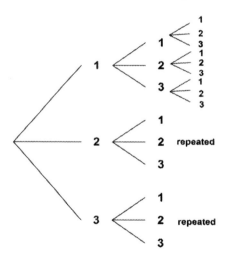

   d. Draw 3 balls three times without replacement

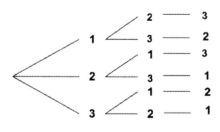

9.   a. Rolling a six-sided die followed by the roll of a four-sided die

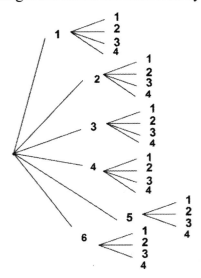

b. Rolling a four-sided die followed by a six-sided die

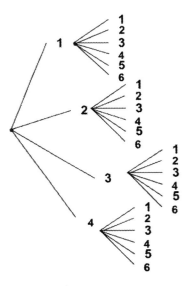

c. The diagrams in parts (a) and (b) illustrate that the first branch has six choices followed by four choices versus four choices followed by six choices. Each has 24 branches at the second step, and the outcomes are the same except for order.

10. a.

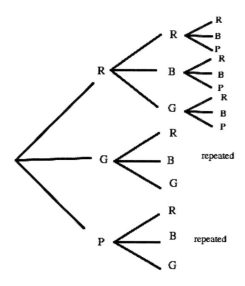

b. Sample space:

RRR	GRR	PRR
RRB	GRB	PRB
RRP	GRP	PRP
RBR	GBR	PBR
RBB	GBB	PBB
RBP	GBP	PBP
RGR	GGR	PGR
RGB	GGB	PGB
RGP	GGP	PGP

c. P(at least one purple) $= \frac{1}{36} + \frac{1}{18} + \frac{1}{12} + \frac{1}{36} + \frac{1}{18} + \frac{1}{12} + P$(last column of sample space).

P(at least one purple) $= \frac{2}{3}$. Alternatively,

$$P\text{(at least one purple)} = 1 - P\text{(no purple)} = 1 - (\tfrac{2}{3} \times 1 \times \tfrac{1}{2}) = \tfrac{2}{3}.$$

d. P(no red) $= P$(GBB) $+ P$(GBP) $+ P$(GGB) $+ P$(GGP) $+ P$(PBB) $+ P$(PBP) $+ P$(PGB) $+$
P(PGP). P(no red) $= \frac{11}{24}$ Alternatively, P(no red) $= \frac{2}{3} \times \frac{11}{12} \times \frac{3}{4} = \frac{11}{24}$.

11. First there are 35 branches, each of these has 34 branches, each of these has 33 branches,
each of these has 32 branches, and finally each has 31 branches.

12. Students might have said that if a marble is taken from the bag without looking, it is most
likely to be yellow because the fraction $\frac{1}{12}$ might *seem* like the biggest.

Answers for the 28.2–28.3 Supplementary Learning Exercises follow the answers for 28.3.

Supplementary Learning Exercises for Sections 28.2 and 28.3

1. a. 1H   1T
        2H   2T
        3H   3T
        4H   4T

    b. $P(2\text{H}) = \frac{1}{8}$

    c. $P(\text{an even } and \text{ H}) = \frac{1}{4}$

    d. $P(3) = \frac{1}{4}$

2. Make a tree diagram if needed.

    a. $P(\text{same color both times}) = P(\text{RR}) + P(\text{GG}) + P(\text{BB}) + P(\text{YY}).$

       $P(\text{same color both times}) = \frac{1}{12} + \frac{1}{12} + \frac{1}{24} + \frac{1}{24} = \frac{6}{24}$

    b. $P(\text{RY}) = \frac{1}{24}$

    c. $P(\text{YR}) = \frac{1}{12}$

    d. $P(\text{they are the same color } or \text{ exactly one is yellow}) = P(\text{same color}) + P(\text{exactly one is yellow}).$
       $P(\text{exactly one is yellow}) = P(\text{RY}) + P(\text{GY}) + P(\text{BY}) + P(\text{YR}) + P(\text{YG}) + P(\text{YB}) = \frac{8}{24}.$
       Therefore, $P(\text{same color } or \text{ exactly one is yellow}) = \frac{6}{24} + \frac{8}{24} = \frac{14}{24}.$

3. Make a tree diagram if needed. Note that without replacement, the probabilities change in the second set of branches.

    a. YY   BY   GY
       YB   BB   GB
       YG   BG   GG

    b. $P(\text{two different colors}) = 1 - P(\text{same colors})$

       $P(\text{same colors}) = P(\text{YY}) + P(\text{BB}) + P(\text{GG})$

       $P(\text{YY}) = \frac{4}{9} \cdot \frac{3}{8} = \frac{12}{72}$ $\qquad$ $P(\text{BB}) = \frac{2}{9} \cdot \frac{1}{8} = \frac{2}{72}$ $\qquad$ $P(\text{GG}) = \frac{3}{9} \cdot \frac{2}{8} = \frac{6}{72} = \frac{1}{12}$

       So, $P(\text{same colors}) = \frac{5}{18}$, and $P(\text{two different colors}) = 1 - \frac{5}{18} = \frac{13}{18}.$

    c. $P(\text{one ball yellow } and \text{ the other ball blue}) = P(\text{YB}) + P(\text{BY})$

       $P(\text{YB}) = \frac{4}{9} \cdot \frac{2}{8} = \frac{8}{72} = \frac{1}{9}$

       $P(\text{BY}) = \frac{2}{9} \cdot \frac{4}{8} = \frac{1}{9}$ $\quad$ So, $P(\text{one yellow and one blue}) = \frac{2}{9}.$

    d. $P(\text{YB}) = \frac{4}{9} \cdot \frac{2}{8} = \frac{8}{72} = \frac{1}{9}$

4.

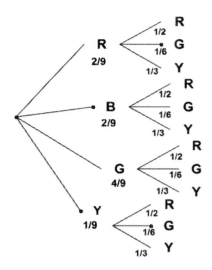

5. a. Without replacement (with a tree diagram you will notice that, surprisingly, the outcomes are equally likely).

RRG	RGR	GRR	YRR
RRY	RYG	GRY	YRG
RGY	RYR	GYR	YGR

b. $P(\text{RRG}) = \frac{1}{12}$

c. $P(\text{GYR}) = \frac{1}{12}$

d. $P(\text{at least two of the same color}) = \frac{6}{12}$

e. $P(\text{all the same}) = 0$

6. a. With replacement (with a tree diagram, you will notice that these outcomes are *not* equally likely).

RRR	GRR	YRR
RRG	GRG	YRG
RRY	GRY	YRY
RGR	GGR	YGR
RGG	GGG	YGG
RGY	GGY	YGY
RYR	GYR	YYR
RYG	GYG	YYG
RYY	GYY	YYY

b. $P(\text{RRG}) = \frac{1}{16}$

c. $P(\text{GYR}) = \frac{1}{32}$

d. Perhaps easier than finding probabilities for all the events in the outcome is

$P(\text{at least two of the same color}) = 1 - P(\text{no two of same color}) = 1 - \frac{6}{32} = 1 - \frac{3}{16} = \frac{13}{16}$,

by looking at RGY, RYG, GRY, GYR, YRG, YGR, each with probability $\frac{1}{32}$.

e. $P(\text{all the same}) = \frac{1}{8} + \frac{1}{64} + \frac{1}{64} = \frac{5}{32}$

f. In this scenario, the probability of choosing a given color does not change from drawing of first ball to drawing of second to drawing of third. The probability of drawing a red is always twice that of drawing a green or a yellow.

7. a.

	1 on first toss	2 on first toss	3 on first toss	4 on first toss	5 on first toss	6 on first toss	7 on first toss	8 on first toss
1	(1,1)	(2,1)	(3,1)	(4,1)	(5,1)	(6,1)	(7,1)	(8,1)
2	(1,2)	(2,2)	(3,2)	(4,2)	(5,2)	(6,2)	(7,2)	(8,2)
3	(1,3)	(2,3)	(3,3)	(4,3)	(5,3)	(6,3)	(7,3)	(8,3)
4	(1,4)	(2,4)	(3,4)	(4,4)	(5,4)	(6,4)	(7,4)	(8,4)
5	(1,5)	(2,5)	(3,5)	(4,5)	(5,5)	(6,5)	(7,5)	(8,5)
6	(1,6)	(2,6)	(3,6)	(4,6)	(5,6)	(6,6)	(7,6)	(8,6)
7	(1,7)	(2,7)	(3,7)	(4,7)	(5,7)	(6,7)	(7,7)	(8,7)
8	(1,8)	(2,8)	(3,8)	(4,8)	(5,8)	(6,8)	(7,8)	(8,8)

b. 64

c. $\frac{32}{64} = \frac{1}{2}$

d. $\frac{16}{64} = \frac{1}{4}$

8. a.

	1	2	3	4	5	6	7	8
1	2	3	4	5	6	7	8	9
2	3	4	5	6	7	8	9	10
3	4	5	6	7	8	9	10	11
4	5	6	7	8	9	10	11	12
5	6	7	8	9	10	11	12	13
6	7	8	9	10	11	12	13	14
7	8	9	10	11	12	13	14	15
8	9	10	11	12	13	14	15	16

b. $\frac{5}{64}$

c. $\frac{3}{64}$

d. The sum of 9 has the greatest probability.

9.   a. $P(X \text{ or } Y) = P(X) + P(Y) - P(X \text{ and } Y) = 0.35 + \frac{7}{40} - \frac{2}{40} = 0.475$
     b. $P(X \text{ or } Y) = 0.35 + \frac{7}{40} = 0.525$

     c. *Disjoint* and *mutually exclusive* mean the same thing, so the probability is the same as in part (b).

     d. Because X and Y are independent,
        $P(X \text{ and } Y) = P(X) \cdot P(Y) = 0.35 \cdot \frac{7}{40} = \frac{49}{800} = 0.06125$.

10. If one *or both* of the parts separated by the "or" hold, the statement is regarded as true.

11. a. $6 \times 6 \times 6 = 216$
     b. $\frac{1}{216}$

     c. X can happen in 6 ways (3-4-1, 3-4-2, 3-4-3,…, 3-4-6), and Y can happen in 36 ways, one of which is 3-4-6. So,
        $P(X \text{ or } Y) = P(X) + P(Y) - P(X \text{ and } Y) = \frac{6}{216} + \frac{36}{216} - \frac{1}{216} = \frac{41}{216}$.

     d. $\frac{1}{216}$ because 3-4-6 is the only outcome in the event X *and* Y.

     e. Yes. The red and white dice must have a total of at least 2 dots. If the green die has 6 dots (Y), that would give a total of at least 8 dots and hence it is not possible to get 7 dots.

     f. $P(Y \text{ and } Z) = 0$

12. a. Not disjoint, because the outcome, king of hearts, would be in both events.
     b. Disjoint, because clubs are black and each card is a single color. So it would not be possible for a card to be both red and a club.

     c. Not disjoint, because 2 would favor both events.

     d. Not disjoint, because some pet owners have both a cat and a dog.

     e. Not disjoint, because most states require both forms.

13. a. Not independent, because potentially the first draw could be the king of hearts, thus influencing the probability of a king on the second draw.
     b. We'll say independent, although you could argue that pitchers will bear down more on the second at-bat, or that having made a hit, the batter will be more confident the second time.

     c. Not independent, for most students. Doing homework should help the probability of getting an A on the test.

     d. Not independent, because a square is automatically a quadrilateral.

14. a. These two events are independent: Getting red on the first spinner, and getting blue on the second spinner. (In fact, any choice of two colors on the separate spinners would give independent events).

   b. These two events are not independent: Getting red on first spinner, and getting red-red. (There are other possibilities.)

   c. These two events are disjoint: Getting red on the first spinner, and getting blue on the first spinner. (There are other possibilities.)

   d. The two events, getting red on at least one spinner, and getting blue on at least one spinner, are not disjoint, because the outcome red-blue favors both events. (There are other possibilities.)

15. a. Events A and B are disjoint.

   b. Event A might involve only outcomes that are in B also. Then

   $P(A) + P(B) - P(A \text{ and } B) = 0.2 + 0.35 - 0.2 = 0.35$.

   c. Not possible. Although $P(A) + P(B) - P(A \text{ and } B) = 0.2 + 0.35 - 0.35 = 0.2$, this would mean that the event (A and B) had *more* outcomes than A itself!

   d. Not possible. Use an argument similar to that of part (c).

   e. Not possible. From $P(A) + P(B) - P(A \text{ and } B) = 0.2 + 0.35 - x = 0.25$, x would equal 0.30, forcing the event (A and B) to have more outcomes than A itself does.

   f. From $P(A) + P(B) - P(A \text{ and } B) = 0.2 + 0.35 - y = 0.4$, y would equal 0.15, and this could happen if the event (A and B) involved only some of the events in A.

Supplementary Learning Exercises for Section 28.4

1. a. The women make up 65% of the 4800 employees, so 3120 of the employees are women. Assuming that the 10% college-diploma rate applies to the women alone, Janine would get 312 from 10% of 3120.

   b. No, Janine's assumption was incorrect. As the following contingency table shows, only 228 of the women have a college diploma. (The Roman numerals show one order of making the calculations, using the given data.)

	Women	Men	Totals
College diploma	vi. 228	v. 252	ii. 480
No coll-diploma		iv. 1428	
	i. 3120	iii. 1680	4800

2. a. There is not enough information, unless you assume that a college diploma is necessary for a management position.

    b. 4% of 4800 = 192 management employees. 6.25% of 192 = 12 management employees without a college diploma, so 192 – 12 = 180 management employees do have a college diploma, leaving 480 – 180 = 300 college diploma holders in nonmanagement positions. Because 96% of the employees, or 4608 of them, are in nonmanagement positions, 300 ÷ 4608 = about 6.5% of the nonmanagement employees have college diplomas. (Yes, this is a complicated exercise! Try a contingency table, with diploma/not and management/not the two headings, to see how the table makes it easier to keep track.)

3.

	Snore	Does Not Snore	Totals
Sleep Apnea	33,100	33,100	66,200
No Sleep Apnea	556,900	376,900	933,800
Totals	590,000	410,000	1,000,000

4.

	Snore	Does Not Snore	Totals
Sleep Apnea	100,000	100,000	200,000
No Sleep Apnea	490,000	310,000	800,000
Totals	590,000	410,000	1,000,000

5. For independent events, what has happened in the past has no influence on what happens in the future.

6. a. Answer a

    b. Because the puppy is brown, it is one of the 4 brown puppies.

7. For independent events A and B, (i) $P(A \text{ and } B) = P(A) \times P(B)$ and (ii) $P(A \mid B) = P(A)$.

8. a. The sample space is reduced by the condition to the 130 who favor removal, so $P(\text{male} \mid \text{favors removal}) = \frac{9}{13}$.

    b. The sample space this time is reduced by the condition to the 120 males, so $P(\text{favors removal} \mid \text{male}) = \frac{9}{12} = \frac{3}{4}$.

    c. There are 90 + 30 + 60 = 180 who are male or oppose removal, so the probability is $\frac{180}{220} = \frac{9}{11}$.

    d. $\dfrac{30}{220} = \dfrac{3}{22}$

9. The bold **500** gives rise to the other values, possibly in the order and manner indicated by the Roman numerals.

	Parents	Non-parents	
Favor bridge	iv. 60 (150 – 90)	v. 175 (= 50% of 350)	vii. 60 + 175 = 235
Favor park	iii. 90 (60% of 150)	vi. 175 (= 350 – 175)	viii. 90 + 175 = 265
	i. 150 (= 30% of 500)	ii. 350 (= 500 – 150)	**500**

a. $100\% - 60\% = 40\%$ (or $\frac{60}{150} = \frac{2}{5} = 40\%$)

b. $\frac{60}{235} = \frac{12}{47} \approx 26\%$

c. $\frac{235+90}{500} = \frac{325}{500} = \frac{65}{100} = 65\%$

d. $\frac{90}{500} = \frac{18}{100} = 18\%$

Answers for Chapter 29: Introduction to Statistics and Sampling

Supplementary Learning Exercises for Section 29.1

1. a. Strictly speaking, a grade of B is not a numerical value. In addition, using just one datum to predict the future is risky.

    b. All right

    c. Assuming that a pattern of statistics will continue is risky.

2. a. Examples: Percent of all students using a computer daily. Percent of all students using a particular website daily. Average number of minutes on a computer per day by computer-using students. Number of students using Macintosh computers.

    b. Examples: Number of beagles in the kennel. Daily number of pounds of food eaten by the dogs in the kennel. Weekly cost of veterinarian care for dogs in the kennel.

    c. Examples: Number of crayons that are not broken. Number of crayons that are shades of red.

3. Answers will vary, of course. Check that your answer refers to the value, perhaps unknown, of a quantity.

Supplementary Learning Exercises for Section 29.2

In 1–4, other answers might be acceptable. If your answer departs considerably from the one given, you might ask someone else to evaluate your answer.

1. Best (as always) is a simple random sample of the students at the university. The sample must be chosen carefully, of course. For example, randomly sampling students close to a parking area would likely involve a bias toward those who drive, whereas sampling students by a bus stop would likely involve a bias toward those who do not drive.

2. If the winning grade is known, the principal could randomly sample children in that grade, perhaps stratified by gender, or even poll all the children in that grade. If the winning grade is not known, the principal could use a random sample stratified by grade (and perhaps stratified within each grade as well).

3. If the producer has a database of its viewers (unlikely), then a random sample of them would be appropriate. If from some source the number of viewers of the news show is known, then the number of "hits" to the website could be used, even though a single user might visit the website multiple times. In the more likely case that the viewers are not known, then telephone calls to a random sample could be used, with the results for those who watch the news show giving the statistic of interest.

4. A few cluster samples (perhaps just one) as the machine operates might show whether the machine is off-kilter. A sytematic sample, say by measuring every 100th container, could be used on an ongoing basis.

5. If the population is quite small and accessible, polling the whole population is reasonable. If the stakes are high, the whole population might be used even if it is large. As an example, the DNA of every inhabitant in an isolated community might be collected to try to identify a serial killer.

6. a. self-selected, or voluntary, sampling from those given the telephone number, even though the initial effort was systematic (the ones who respond)

    b. cluster sampling

    c. systematic sampling

    d. self-selected, or voluntary (the ones who respond)

    e. self-selected, or voluntary (the ones who respond)

7. a. If only Ms. Allen's class is involved, the 60% would be a population parameter. However, if the 60% result is being used to represent all fourth-graders or even all students, the 60% would be a sample statistic.

    b. The 42% is a population parameter, for the population of all students taking Algebra I for the first time in Jefferson School District.

    c. "Survey" implies a sample, so the 89% is a sample statistic.

Supplementary Learning Exercises for Section 29.3

1. One way would be to ignore the digits 0, 7, 8, and 9 and look at consecutive pairs from the table. For example, 184923657 would become 142365, giving "rolls" of (1, 4), (2, 3), and (6, 5).

2. a. One way is to proceed as in Exercise 1 but then to record the sum. For the results in Exercise 1, you have sums 5, 5, and 11. It is important to see that using two-digit numbers to search for the sums 02 through 12 does not take into account the fact that the sums with dice are not equally likely. The sums would incorrectly be equally likely if the table is used to look for sums 02 through 12.

    b. The difference is in the interpretation of the result. For example, (1, 4) is interpreted as 5. Check to see that your method does not give equally likely sums.

    c. With a large number of repetitions, you should get a sum of 7 about $\frac{1}{6}$, or roughly 17%, of the time.

3. Assume that the faces ("sides") of the dice are labeled 1, 2, 3 and 4. The simplest way is to use pairs of digits from the table but ignoring any of the digits 0, 5, 6, 7, 8, and 9, much as in the answer to Exercise 1.

4. a. The list has 12 names. One way would be to put each of the 12 names on identically sized slips of paper, put the slip in an opaque bag (mixing them up), and draw 5 slips (without looking). A TRSD could be used by numbering the 12 names and then using two-digit numbers, ignoring 00, 13 through 99, and any repetitions from 01 through 12 that might occur. Or use five spins of a spinner marked into 12 equal sectors, with the names of the students on the sectors and ignoring repetitions of a name when spinning.

    b. Answers will vary, of course. Were there any surprises in your ten samples? For example, was some one person on a large number of samples?

5. a. Answers will vary.

    b. A TRSD, with even digits being H and odd digits being T, gives one method.

    c. Answer will vary, of course. The interesting thing is that humans usually *avoid* having several H's or T's in a row, but such does happen with random tosses.

Supplementary Learning Exercises for Section 29.4

1. a. categorical (One might count the number of Tom-Dick-Harry finishes, but T-D-H is just a category.)

    b. categorical (If by some strange chance, someone was thinking "length of favorite movie," then it would be measurement.)

    c. measurement, although if answers are confined to choices such as "lots," "average," "little," and "none," the data would be categorical. Some books would label these categories as *ordinal* data because there is a natural order to the categories.

    d. measurement, if the focus is the number of books; categorical, if the focus is the particular titles

    e. measurement, if the focus is the number of children; categorical, if birth orders like GBG for 3-children families is the focus

2. Samples are given here. There may be other possibilities.

    a. measurement: number of hours or minutes; categorical: types of sports

    b. measurement: number of laps swum, number of competitors, or time for race; categorical: type of stroke, length of race

    c. measurement: amount spent, number of stores visited, or number of items purchased; categorical: types of items sought or types of stores visited

    d. measurement: value of prizes, loudness (singer or audience), or number of votes; categorical: types of songs

3. The numbers just describe the different categories, so they give categorical data.

Answers for Chapter 30: Representing and Interpreting Data with One Variable

Supplementary Learning Exercises for Section 30.1

1. a–b. Following are graphs from Excel; compare yours for accuracy. The percents are calculated by dividing the travel mode by the total number of children, 150; angle sizes are calculated in this way:

   For bus travel, $\frac{90}{150} \cdot 360 = \frac{3}{5} \cdot 360 = 3 \cdot 72 = 216$ degrees.

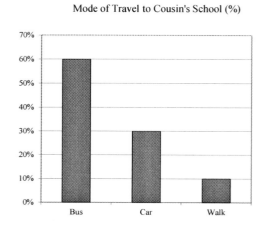

Mode of Travel to Cousin's School (%)

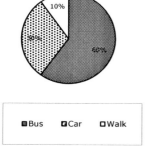

Mode of Travel to Cousin's School

   c. Your pictograph should have 6 of your basic symbols for bus, 3 for car, and 1 for walk.

   d. Because of the larger percent by bus and the relatively small percent by walking for the cousin's data than for Jasmine's school, the cousin's school is the one in a rural area.

2. The graphs for Exercise 1 above are from Excel.

3. a. Many of the values are decimals, so showing parts of some basic symbol could be messy.

   b. The sum of the values does not make sense here.

c. Here is a graph from Excel.

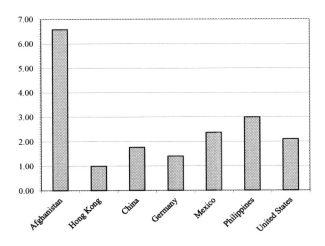

Fertility Rates in Selected Countries

d. Low fertility rates might result in a shrinking population, with an increasing percent of older people. A high fertility rate might result in a high rate of increase in population, with accompanying strains on resources. (Different immigration and emigration figures, as well as child mortality rates, are of course potential influences on population. Hong Kong, for example, might have large numbers of immigrants, and Afghanistan might have a high child mortality rate.)

4. a. (from Excel)

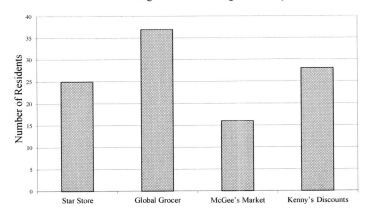

Neighborhood Patronage at Grocery Stores

b. (from Excel)

Neighborhood Patronage at Grocery Stores

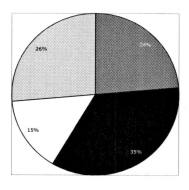

c. In the bar graph it is easy to see a difference in height. Without the percent labels it is difficult to tell the exact portions on the circle graph.

5. Here is a slightly different style from the one in the answer for Exercise 4.

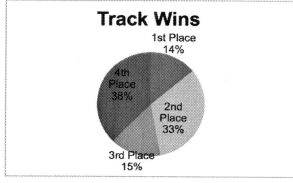

6.

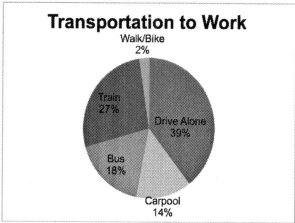

7. Questions will vary. Here is one: What assumptions does this graph make? (A student does only one thing. All students participate.)

Supplementary Learning Exercises for Section 30.2

1. a–b. Here are a sample stem-and-leaf plot and (from Fathom) a histogram for the scores.

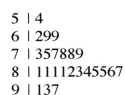

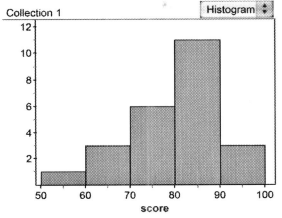

c. The stem-and-leaf plot allows you to tell each individual score, whereas the histogram does not tell where the individual scores are located within the bar.

2. a. i. 7 rows 3–9   ii. 10 rows (or 11)   iii. 56 rows, 39 through 94
   iv. The last digit does not indicate a significant amount relative to the count.

   b. The stem plot could vary depending on your answers in part (a). Truncating (omitting) the ones digit gives the following plot.

   The stem is in the hundreds, so 3 | 9 9 represents 390 and 395.

   ```
   3 | 9 9
   4 | 0 1 1 2 3 3 4 5 7 8 9
   5 | 2 4 4 5 7 8 8 9
   6 | 5
   7 | 0 5 8
   8 |
   9 | 1 4
   ```

   c. The data are mostly bunched between 400 and 600.

   d. 523

3. a. Circle graph, bar graph

   b. Stem-and-leaf plot (histogram, box plot from the next section)

   c. Circle graph, bar graph

    d. Stem-and-leaf plot (histogram, box plot from the next section)

    e. Circle graph, bar graph

    f. Circle graph, bar graph

4.

2	1 2 4 4 4 4 6 6 6 6 6 6 7 7 7 7 8 8 9 9 9
3	0 0 0 0 0 1 1 1 2 3 3 3 3 4 4 4 4 4 5 5 5 7 7 7 7 8 8 8 9 9 9
4	1 1 1 1 2 2 5 5 7 9 9
5	
6	0 1 1 2
7	4
8	1

5. Your graph should carry the same ideas as the one given here. The colors may be difficult to distinguish in this version. Notice the clutter, even without percents.

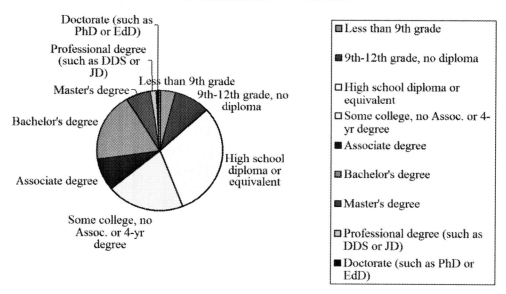

Educational Attainment of 18-64 Year Olds

Supplementary Learning Exercises for Section 30.3

1. a. The data would be categorical (for example, chocolate, strawberry), so a five-number summary does not make sense.

    b. This situation would give measurement data, so a five-number summary makes sense.

    c. This situation would give measurement data, so a five-number summary makes sense.

    d. The data would be categorical, so a five-number summary does not make sense.

    e. This situation would give measurement data, so a five-number summary makes sense.

2. Neither Arnold's nor Bonnie's parents should be concerned about the child failing, because each scores in the upper half of all the scores. Carl's parents, on the other hand, might worry, because 52% on a semester test is not a good score at all.

3. a. Ignoring the two outliers at 53 and 62, the five-number summary is 68 (low), 78 (first quartile score), about 81 (median), about 88 (third quartile score), 99 (high score). The second quartile involves several scores bunched between 78 and about 81.

    b. This box plot does not contain information about the number of scores, nor about the individual scores. (However, such information is available with the software.)

4. This is not statistically possible. Half of the scores will be below the median by definition. The median is dependent on the scores of the test and is not a value that is predetermined. The district could possibly set a goal for passing, and this could be a specific score that is not dependent on the outcomes.

5. a. Here is a sample box plot (from Fathom).

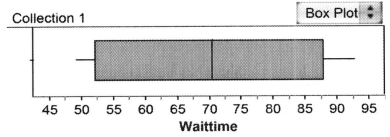

    Your box plot should involve (roughly) this five-number summary:

Minimum	Q1	Median	Q3	Maximum
49	52	70.5	88	93

    b. In this case, we no longer have the order in which they occurred, nor any knowledge about their exact values (except for the minimum and maximum).

6. Your answers may differ from the samples given, of course.

    a. **14**, 15, **18**, 19, 19, 20, 20, **22**, 32, **40**

    b. **82**, 82, 82, 82, **82**, **82**, 94, **94**, 98, **98**

    c. **10**, 10, **19**, 19, **26, 28**, 48, **59**, 60, **75**

7. Parts of your box plots will depend on your answers, of course. The following Fathom plots are for the data sets given in the answer to Exercise 6.

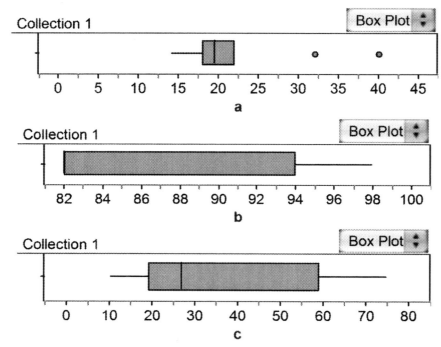

8. Your box plot should look something like this one from Fathom.

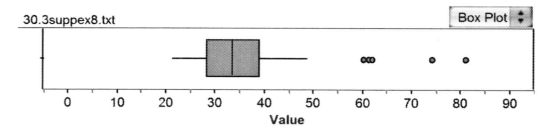

9. a. Histogram a goes with the third box plot.

    b. Histogram b goes with the middle box plot.

    c. Histogram c goes with the first box plot.

10. a. 58, 72, 76, 86, 95   b. 14, 17.5, 20.5, 25, 32

11. $0.78 \times 485 \approx 378$ students scored equal to or lower than Roger.

12. There are roughly 35% of the cases between the 30th and 65th percentiles. So there are roughly $0.35 \times 750 \approx 262$ students between the 30th and 65th percentiles.

Supplementary Learning Exercises for Section 30.4

1. a. Median. There could be outliers of high salaried management.

    b. Mean. Prices tend to be similar in a region.

    c. Mode. Types of car owned give categorical data.

    d. Mean, unless there are some extreme outliers.

    e. Median. Some houses can have extremely high prices compared to others in the area.

    f. Mean, although mode might fit kindergarteners better.

2.   a. Mean = 11   median = 7.5

    b. Mean = 8.45   median = 7 The mean is affected more than the median by the relatively large 39 (for 12 cases). When the 39 is removed (now 11 cases) the median is moved by at most one case.

3.   a. Mean = 4.85   standard deviation = 0.286 feet (0.279 on some calculators)

    b. Mean = 58.25 standard deviation = 3.432 inches (3.345 on some calculators)

    c. The numbers have increased by a factor of 12, but the actual values (with units) are the same. For example, 0.286 feet = 3.432 inches.

    d. Your graphs should have the same shape but different scales.

4. $3 + 4 + 4 + 6 = 17$ dozen, or $17 \times 12 = 204$, cupcakes. With 4 nights involved, that gives an average of 51 cupcakes per night.

5.   a. Possible

    b. Possible (if each cat weighed 14 pounds)

    c. Not possible (Why?)

    d. Possible

    e. Not possible (Why?)

    f. Possible (if the third cat weighs enough more than 14 pounds to make up for the lighter cats)

6. $(32 \times 8.50) + (8 \times 12.75) = 374$ dollars for the 40 hours, an average of $9.35 per hour

Supplementary Learning Exercises for Section 30.5

1.   a. First reorder the scores: 2, 5, 8, 9, 9, 12, 13, 16, 19, 19, 21, 24. The range is $24 - 2 = 22$.

    b. There are 12 values in the set, so the median is the average of the two middle scores, 12 and 13, which is 12.5. There are 6 values in the bottom half of the set, and the median of the bottom half is the average of 8 and 9, which is 8.5. So 8.5 is the first quartile. The median of the top half is 19, which is the average of 19 and 19. So 19 is the third quartile. The five-number summary is 2, 8.5, 12.5, 19, 24.

c. The interquartile range is $19 - 8.5 = 10.5$. Outliers would need to be less than $8.5 - 15.75 = {}^-7.25$ or greater than $19 + 15.75 = 34.75$. There are no such values among the scores ($1.5 \times 10.5 = 15.75$).

d. This box plot is from *Illuminations*, but you should be able to draw one of your own. Remember to keep the units equal in size on the bottom axis.

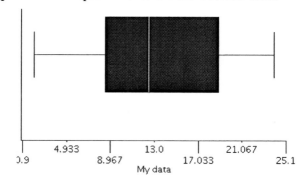

e. The mean of the scores is slightly more than 13.

f. The distribution is bimodal. The two modes are 9 and 19.

g. The mean (13) and the median (12.5) are quite close for these data. Because there is not one mode, the mode could not be compared to the mean and the median.

h. The average deviation is 5.6. The calculation is shown below.

Values	Distance from 13	Squared distance from the mean
2	11	121
5	8	64
8	5	25
9	4	16
9	4	16
12	1	1
13	0	0
16	3	9
19	6	36
19	6	36
21	8	64
24	11	121
Sum	**67**	**509**

$67 \div 12 = 5.6$ $509 \div 12 = 42.4$ is the variance, so

$$\sqrt{42.4} \approx 6.5 \text{ is the standard deviation.}$$

i. The standard deviation is 6.5. The calculation is shown above.

j. 5.6 units and 6.5 units are fairly close. Both numbers provide information about the spread of the data. The five-number summary is also an indication of spread, but of a different type.

2. a. The average *yearly* rainfall for New Orleans is 61.88, and for San Diego it is 9.90. (Just add the monthly averages.)

b. The average *monthly* rainfall for New Orleans is 5.16, and for San Diego it is 0.83. (Find the average of the monthly rainfall amounts.)

c. Answers will vary. Recall that the standard deviation is a measure of spread.

d. The standard deviation of monthly rainfall for New Orleans is 0.888 (or 0.928 on some calculators), and for San Diego it is 0.706 (or 0.738 on some calculators). Why are the standard deviations so close while the means are, relatively speaking, so far apart?

Supplementary Learning Exercises for Section 30.6

1. a. Distribution *B*. The high point of a normal curve corresponds to the mean on the horizontal axis, and the high point of graph *B*, corresponding to 90, is farther to the right than is that of graph *A*, corresponding to 75.

b. Distribution *B*. The high point of a normal curve corresponds to the median on the horizontal axis, and the high point of graph *B* corresponding to 90, is farther to the right than is that of graph *A*, corresponding to 75.

c. Distribution *B*. The high point of a normal curve corresponds to the mode on the horizontal axis, and the high point of graph *B*, corresponding to 90, is farther to the right than is that of graph *A*, corresponding to 75.

d. Distribution *B*. The more spread out a normal curve is, the greater the standard deviation, and graph *B* is more spread out.

e. Distribution *B*. The variance is the square of the standard deviation, so the greater standard deviation, part (d), will give the greater variance.

f. Distribution *A*. The *z*-score tells how far you are above or below the mean (in standard deviation units). In distribution *A*, *x* is above the mean (75) and so will give a positive *z*-score, but in distribution *B*, *x* is below the mean (90) and would have a negative *z*-score.

2. Ann's and Beth's actual scores are the same distance from the mean. Ann's is 0.8 standard deviation below the mean, and Beth's is 0.8 standard deviation above the mean.

3. a. Answers will vary. Most data will be in the middle and to the left with only a few to the right.

    b. Answers will vary. Most data will be in the middle and to the right with only a few to the left.

4. a. Mean = 4.85   standard deviation = 0.286 feet (0.279 on some calculators)

    b. Mean = 58.25   standard deviation = 3.432 inches (3.345 on some calculators)

    c. For the height in feet: $z = \frac{5.25-4.85}{0.286} \approx 1.4$. For the height in inches: $z = \frac{63-58.25}{3.432} \approx 1.4$ (No surprise, because all the values involved in the computation are multiplied by 12.)

    d. No, the bar graph is not shaped like a bell curve at all.

5. You can find the z value, but it may not be meaningful in terms of the 68-95-99.7 rule. It might be useful in comparing performance on another test or measurement of some sort.

6. a. ⁻0.33   b. 1.53   c. 2.4   d. ⁻0.73   e. ⁻2.33

7. a. Symmetric, normal

    b. Symmetric, non-normal, bimodal

    c. Skewed right, non-normal

8. a. Not possible. A skewed graph has a different median than mean.

    b. Not possible.

    c. Not possible.

    d. A bimodal distribution is an example.

9. Locations are approximate.

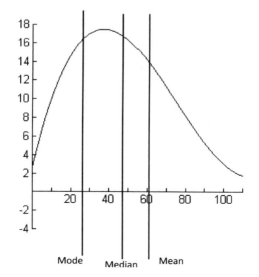

10. The mean would become 89 and the median 83. But the standard deviation would stay the same.

Answers for Chapter 31: Dealing with Multiple Data Sets or with Multiple Variables

Supplementary Learning Exercises for Section 31.1

1. a.

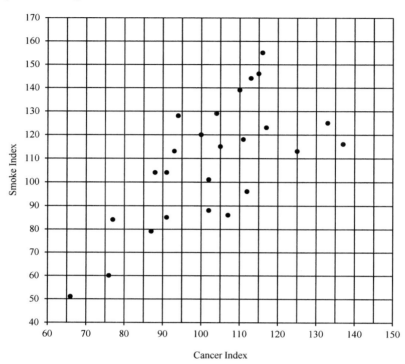

b. The data show a positive association.

c. As the smoking index increases, so does the cancer index. We cannot say that this is a causal relationship without more information, however.

2. a. Type refers to the type of experimental sampling—random in this case.

b. The two variables are size of diamond in carats and price in Singapore dollars.

c. The most expensive diamond cost 1086 Singapore dollars and was 0.35 carats.

d.

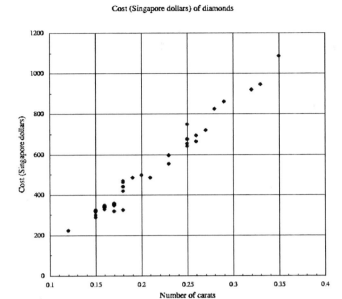

Cost (Singapore dollars) of diamonds

e. The variables are positively associated. This means that the greater the carat size the more the diamond costs.

3.   a. Positively associated, at least until the fish is fully grown

b. Arguments could be made for different answers, but most likely they are positively associated.

c. No association

d. No association

Supplementary Learning Exercises for Section 31.2

1. a.

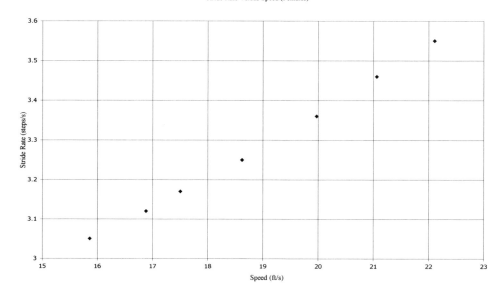

b. These data are very close to being on the line. The *R* value for this graph shows a strong positive association.

c.

Speed (ft/s)	15.86	16.88	17.5	18.62	19.97	21.06	22.11
Stride rate	3.04	3.12	3.17	3.26	3.37	3.45	3.54

2.   a. close to 1   b. close to ⁻1

    c. close to 0   d. close to 0

3.   a. The graph in 2(c)

    b. The graph in 2(d)

    c. The graph in 2(a)

    d. The graph in 2(b)

4.   a. Close to 1   b. Close to 1, but not as close as in part (a)

    c. Close to 0   d. Close to ⁻1

5.   The *x*-values from zero to four appear to have a strong negative association. The values from four to ten appear to have a strong positive association. These appear to cancel out, making the overall association closer to 0. This example shows the importance of an awareness of what values of the independent variable are used for a correlation coefficient.

Answers for Chapter 32: Variability in Samples

Supplementary Learning Exercises for Section 32.1

1. a. **36** of the 1000 samples of size 100 gave a percent *against* of 59%; **25** of the 1000 samples of size 500 gave a percent *against* of 59%; **6** of the 1000 samples of size 1000 gave a percent *against* of 59%; and **1** of the 1000 samples of size 2000 gave a percent *against* of 59%.

   b. 1 of the 1000 samples of size 100 gave a percent *against* of 83%.

2. The column should have fewer entries and be condensed even more around the 63% row.

3. The $\frac{1}{\sqrt{n}}$ rule of thumb can be used if the (usually unknown) population parameter is around 50%. The *n* refers to the sample size. The rule gives an idea of the margin of error.

4. 88% suggests that the underlying population parameter is not around 50%. The large sample size strengthens that suggestion.

5. a. True

   b. False! Sample statistics can vary a lot from sample to sample, as Table 1 shows.

   c. True

6.

a.   Number of 1000 Samples (size 1000) with Percents

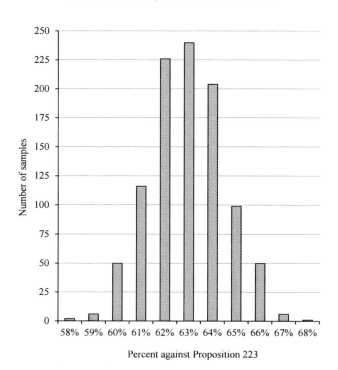

Percent against Proposition 223

b. To a good degree. "Perfect" data would appear to fit a normal curve.

Supplementary Learning Exercises for Section 32.2

1.   a. Jeff's polling of the entire student body (if practical) should give no margin of error. From the $\frac{1}{\sqrt{n}}$ rule of thumb, Angie's idea would give a margin of error of $\pm10\%$, and Jan's would give a margin of error of about $\pm3\%$.

   b. Strictly speaking, *Yes* did win a majority. But the vote was virtually tied, so the university would have to evaluate which group to offend.

   c. Not with great confidence, because the confidence interval is 46%–66%.

   d. With a confidence interval of 51%–57%, the university should be somewhat safe in proceeding.

2.   a. The confidence interval, 48%–56%, does include less than 50%.
   b. From $\frac{1}{\sqrt{n}} \approx 0.04 = \frac{1}{25}$, the sample size was around $25^2 = 625$.

3.   a. 96%–100%
   b. 95%

c. Yes, for a couple of reasons. First, there is only a 95% confidence that repeating such tests would give intervals around the population parameter. Second, the test itself does not appear to be 100% effective.

4. First, the sample of 16 may be biased, because they were all Robin's friends. Second, the confidence interval is large: 54%±25%, or 29%–79%, so the population parameter could well be below 50%.

5. 51% and ± 3%

Answers for Chapter 33: Special Topics in Probability

Supplementary Learning Exercises for Section 33.1

1. The expected value is $(0.1 \cdot 1) + (0.1 \cdot 2) + (0.1 \cdot 3) + (0.1 \cdot 4) + (0.1 \cdot 5) + (0.5 \cdot 6) = 4.5$ spaces.

2. You can roll a 7 six different ways, so the probability of rolling a 7 is $\frac{1}{6}$. From your viewpoint, the cost to play is $^-1$. Hence, the expected value is $\frac{1}{6} \cdot (5 - 1) + (\frac{5}{6} \cdot {}^-1) = {}^-\frac{1}{6}$, and in the long run the owner (not you) will average $\$\frac{1}{6}$ per game.

3. $P(\$0.50) = \frac{1}{2}$, $P(\$0.75) = \frac{2}{6}$, and $P(\$2.00) = \frac{1}{6}$. The expected value, without the charge to play, is $(\frac{1}{2} \cdot 0.50) + (\frac{1}{3} \cdot 0.75) + (\frac{1}{6} \cdot 2.00) = 0.83\frac{1}{3}$ dollars or $83\frac{1}{3}$¢. The owner charges $\$1$ so the owner can expect to make money if the game is played many times.

4. a. The expected value is
    $$(0.05 \cdot 0) + (0.15 \cdot 1) + (0.25 \cdot 2) + (0.30 \cdot 3) + (0.10 \cdot 4) + (0.15 \cdot 5) = 2.7$$
    b. She can expect to get an average of 2.7 calls per day, over the long run.

5. The expected value (from the booth manager's viewpoint) without taking into account the cost to play is $(0.3 \cdot {}^-75) + (0.7 \cdot 0) = {}^-22.5$. The booth should charge at least one 25¢ ticket to play in order to make money.

Supplementary Learning Exercises for Section 33.2

1. $8 \cdot 10 \cdot 10 \cdot 10 \cdot 10 \cdot 10 = 8,000,000$.

2. It depends. Although there are 400 possible codes ($4 \cdot 10 \cdot 10$), you may have 100 or more books in a particular category. It would be safer to use more digits to extend your code. If every category of books includes fewer than 100 books, then you are safe.

3. $20 \cdot 19 \cdot 18 \cdot 17 \cdot 16 \cdot 15 = 27,907,200$ different orders! If all you care about is the choices of CDs, without concern about the order in which they're placed, there are "only" $_{20}C_6 = 38,760$ choices.

4. Taking into account the number of ways each part of the committee can be chosen, and the fact that every part can go with each of the other parts, there are
    $2 \cdot 3 \cdot {}_6C_4 \cdot {}_9C_4 = 2 \cdot 3 \cdot 15 \cdot 126 = 11,340$ ways in which the committee can be formed.

5. $13 \cdot {}_{15}C_2 \cdot 9 = 13 \cdot 105 \cdot 9 = 12,285$ possible different committees

6. There are $_{52}C_5$ different 5-card hands. The three face cards can occur in $_{12}C_3$ ways, and the other two cards, which must come from the 40 non–face cards, can occur in $_{40}C_2$ ways. The probability, then, is $\dfrac{{}_{12}C_3 \cdot {}_{40}C_2}{{}_{52}C_5} = \dfrac{220 \cdot 780}{2,598,960} \approx 0.066$.

7.   a. $10 \cdot 26 \cdot 26 \cdot 26 \cdot 10 \cdot 10 = 17,576,000$, although some of the 3-letter choices might not be allowed.

   b. $10 \cdot 26 \cdot 25 \cdot 24 \cdot 9 \cdot 8 = 11,232,000$ but again some of the 3-letter choices might not be allowed.

8.   a. $_8C_4 = 70$

   b. $\dfrac{_5C_3}{_8C_3} = \dfrac{10}{56} = \dfrac{5}{28} \approx 18\%$

   c. $\dfrac{_5C_3 \cdot _3C_2}{_8C_5} = \dfrac{30}{56} = \dfrac{15}{28} \approx 54\%$

9.   Two balls can be chosen in $_9C_2 = \frac{9 \cdot 8}{2 \cdot 1} = 36$ possible outcomes. This problem is complicated because there are so many ways that you might get \$10 or more.

   Draw two \$5 balls from the four \$5 balls: $_4C_2 = 6$ outcomes.
   Draw a \$10 ball (2 ways) and one of the other 8 balls: $2 \times 8 = 16$ outcomes.
   So the probability of drawing at least \$10 is $\frac{22}{36} = \frac{11}{18}$.